DECISION AND DISCRETE MATHEMATICS
maths for decision-making in business and industry

Mathematics possesses not only truth, but supreme beauty - a beauty
cold and austere, like that of sculpture, and capable of stern perfection,
such as only great art can show.

> Bertrand Russell in *The Principles of Mathematics*

Etymologia Scientifica

Algorism (æ·lgŏriz'm). ME. [a. OFr. *au-
gorisme, algorisme,* ad. med.L. *algorismus,* f.
Arab. *al-Khowarazmi,* i.e. *native of Khiva,* sur-
name of an Arab. mathematician. Cf. *Euclid =*
plane geometry.] The Arabic, or decimal system of numeration ; *hence,* arithmetic. Also
attrib.
     Corruptlye written..Augrim for algorisme, as the
Arabians sounde it RECORDE.   Hence **Algori·smic**
a. arithmetical.
Algorithm, erron. refash. of ALGORISM.

(Shorter Oxford English Dictionary on Historical Principles)

About the Spode Group

The Spode Group was formed in 1980 by a number of dedicated teachers of mathematics who were resolved to improve the standard of mathematics teaching and learning throughout schools and universities. They planned to produce relevant and stimulating mathematical material for use in teaching, in order to demonstrate how mathematics can be used to solve practical problems in real life situations. As well as producing a wide range of mathematics curriculum material, which included a strong element of software, they initiated such projects as mathematical modelling for teachers, project and practical work in statistics, computing, and course work for the General Certificate of School Education (GCSE).

They met regularly in their "own time" in Exeter and other places for weekend seminars. The Spode Group was then unknown and not acknowledged by the authorities. Their only reward was that of knowing they were improving the teaching standards of their subject, recognising the importance that mathematics plays in the lives of their students and everyone throughout their working and social life.

This small nucleus of enthusiasts, all of whom have now moved to higher academic responsibility, was at that time directed by Dr.David Burghes now Professor of Education at Exeter University and Director of the Centre for Innovation in Mathematics Teaching, in close rapport with Dr John Berry now Professor of Mathematics in Plymouth University, and Dr Ian Huntley now Director of Continuing Education in The University of Bristol, as Associated Directors. With great foresight they envisaged and initiated a new module for mathematics teaching closely related to working and business life, as distinct from the pure or the applied mathematics papers of the A level syllabuses..

In due course they collaborated to write *Decision Mathematics*, the forerunner of our present book which became the recommended set text for the Oxford University Delegacy of Local Examinations for the Advanced Level mathematics option. Published in 1986 by Ellis Horwood Limited the text was most successful, widely used in secondary schools, but is now "out of print". Since that early breakthrough, teaching mathematics for business life has become nationally accepted by all A-level examination boards. This second edition, rewritten, updated and retitled *Decision and Discrete Mathematics*, again provides a coverage for the Discrete Mathematics module.

The Spode Group has since gained support and is widely acknowledged for its significant contribution to mathematics teaching. As the publisher privileged to launch both versions of this mathematical dichotomy in two successive editions, I have enjoyed the friendship of members of the Spode Group throughout all those years. I am sensitive of their trust and support and of our continuing association of Albion Publishing Limited as their publishers, for which honour I shall always be grateful.

Ellis Horwood, the publisher at Albion Publishing
Chichester   June

1996

DECISION AND DISCRETE MATHEMATICS
maths for decision-making in business and industry

Written on behalf of **The Spode Group** by
Ian Hardwick, MA (Oxon)
Head of Department of Mathematics
Truro School
Cornwall

Edited by
Nigel Price, BSc (Aston), MPhil (Exeter)
Innovative Mathematics Teaching Centre
University of Exeter School of Education
Exeter

Albion Publishing
Chichester

First published in 1996 by
ALBION PUBLISHING LIMITED
International Publishers, Coll House, Westergate, Chichester, West Sussex,
PO20 6QL England

© The Spode Group, 1996

British Library Cataloguing in Publication Data
A catalogue record of this book is available from the British Library

ISBN 1-898563-27-6

Printed in Great Britain by Hartnolls, Bodmin, Cornwall

Contents

Preface

Recent decades have seen a vast increase in the developments and applications of mathematics to solve problems requiring discrete mathematics.

The process of solving such problems is often referred to as **operational research**, and employs techniques that have been developed to solve particular classes of problems, often related to the efficient use of resources.

A second class of problems has developed out of the need for solving problems related to the IT revolution – many of these problems require mathematics in their solution, but it is a rather different type of mathematics from the familiar continuous theories.

Both classes of problem have, at their heart, the use of **decision** (or **discrete**) mathematics, and this is the focus of this text.

Many exam boards, at A-level, have now included topics in Discrete Maths in their modular courses. This text, based on the early Oxford Board AS syllabus, has been expanded, and will be of interest to students (and teachers) taking Discrete Maths courses for any exam board.

This totally rewritten version of the original Spode Group text on 'Decision Mathematics' has been written by Ian Hardwick and edited by Nigel Price. We are grateful for their dedication and hard work, and delighted that the new version of our book is now available.

David Burghes

on behalf of **The Spode Group**

Acknowledgements

We are grateful for help with this book to

> Mark Carroll
>
> Joe Chan
>
> Simon Collinge
>
> Ben Ferrett
>
> James Knight
>
> Jo Mooney

and delighted that Ellis Horwood is again publishing our text.

Finally, we are most grateful to Ann Tylisczuk for typing a first draft and to Liz Holland for typesetting the final version.

1

An introduction to networks

1.1 TERMINOLOGY

A **network** is a set of points and lines each of which has its ends at a point or points. Various equivalent terms exist and may be met; these are

network	≡	graph		
node	≡	vertex	≡	point
arc	≡	edge	≡	line
region	≡	face	≡	area

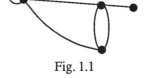

Fig. 1.1

1.2 INVESTIGATIONS

Investigation 1

A point where lines meet is called a **node**. The **order** of the node is the number of ends of lines that meet there.

Draw, if possible, networks with the specifications given in the table.

3 node   6 node

	1 node	3 nodes	4 nodes	5 nodes
(a)	0	2	1	0
(b)	1	1	1	0
(c)	1	2	2	0
(d)	3	3	1	0
(e)	0	3	1	1
(f)	1	1	0	1

Can you give a rule to describe those networks that cannot be drawn?

The order of the node may also be referred to as the **vertex degree** or **valency**.

Investigation 2

A network is said to be **traversable** if it can be drawn with a single stroke of a pen without going over any arc twice or lifting the pen off the paper.

Which of the following are traversable?

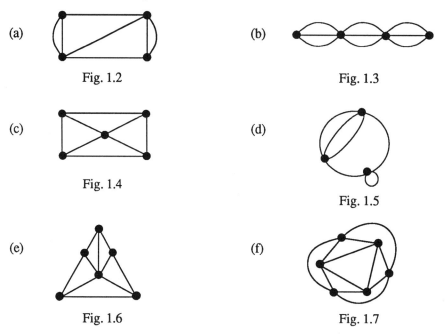

(a)

Fig. 1.2

(b)

Fig. 1.3

(c)

Fig. 1.4

(d)

Fig. 1.5

(e)

Fig. 1.6

(f)

Fig. 1.7

Draw some other networks of your own. By looking at the numbers of odd and even nodes can you find a rule to describe those networks that are traversable?

If you have found a rule, can you explain it?

Investigation 3

A **connected** network is one in which it is possible to get from any of its nodes to any other by a route along its arcs. A **tree** is a connected network so that there is exactly one route between any pair of nodes (i.e. it contains no circuits, where a **circuit** is a route which passes through one or more nodes and finishes at the starting node).

There are only two different trees on four nodes.

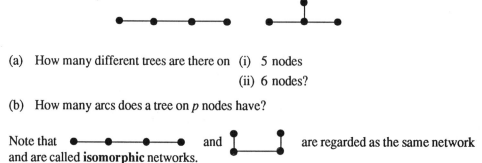

(a)   How many different trees are there on   (i)   5 nodes

(ii) 6 nodes?

(b)   How many arcs does a tree on p nodes have?

Note that ●———●———●———● and ⌐_⌐ are regarded as the same network and are called **isomorphic** networks.

Exercise 2.1 (The Königsberg Problem)   In the fifteenth century the East Prussian city of Königsberg (now known as Kaliningrad, and part of Russia) was divided into four parts by the river Pregal, the parts being joined together by seven bridges. Can you find a route by which the citizens could walk to each of the four parts of the city, crossing each bridge only once and returning to their starting point? (Only four of the original bridges exist today.)

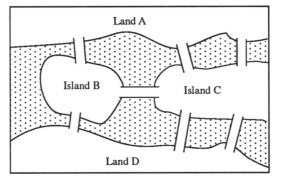

Fig. 1.8

Exercise 2.2   In chemistry, the hydrocarbons are made up of hydrogen and carbon atoms only. Hydrogen atoms may be thought of as one nodes and carbon atoms as four nodes.

Hydrogen   ●——   Carbon   —┼—

An example of a hydrocarbon, ethane, is represented in Fig. 1.9.

Ethane
$C_2 H_6$

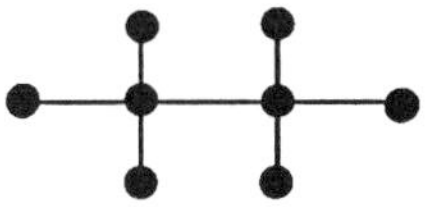

Fig. 1.9

Which of the following could not possibly exist?

(a) $C_3 H_7$ (b) $C_4 H_{10}$ (c) $C_{10} H_{11}$

Nitrogen can be thought of as a three node and oxygen as a two node, so which of the following are impossible?

(d) NH_3 (e) CO (f) CO_2 (g) H_2O (h) N_3O

(i) N_2H_3 (j) C_2NH_2 (k) $C_6H_{12}O_6$ (l) CN_3OH_2?

1.3   MINIMUM CONNECTOR PROBLEM
(MINIMUM SPANNING TREE)

This is the type of problem which involves, for example, finding the least amount of cable needed to connect the towns in the network in Fig. 1.10. If the length used is to be a minimum, then the arrangement must be a tree or it will contain a circuit and therefore an arc that could be omitted.

You need an **algorithm**, that is a process or set of rules, to solve the problem. In this case there is more than one algorithm available to do the job.

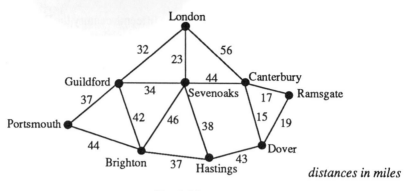

distances in miles

Fig. 1.10

Kruskal's algorithm 1

(a) Start with the complete network and delete arcs according to the following rule.

(b) Delete the arc with the highest number which does not disconnect the network.

(c) Repeat (b) until any deletion disconnects the network.

Kruskal's algorithm 2

(a) Start with the set of nodes alone and add arcs according to the following rule.

(b) Add the arc with the lowest number which does not create a circuit.

(c) Repeat (b) until any addition creates a circuit or until the number of arcs is one less that the number of nodes.

Prim's algorithm

(a) Start with the set of nodes and select any one to consider. Add arcs according to the following rule.

(b) Add the lowest arc that connects a new node to the set under consideration.

(c) Repeat until all the nodes are connected or until the number of arcs is one less than the number of nodes.

In each of these steps, if there is a choice of arcs with equal numbers at any stage, any selection from these will do.

Example

Use the three algorithms to find the minimum spanning tree for the network in Fig. 1.11.

Make three copies and show the orders in which arcs were selected according to the different rules.

Solution

These are examples of **greedy** algorithms, so called because they always take the greediest choice available. Prim's algorithm is usually considered the best on which to base a

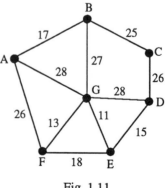

Fig. 1.11

computer program as it can be illustrated by writing the facts of the network as a route matrix and applying the following rules.

	A	B	C	D	E	F	G
A	–	17	–	–	–	–	28
B	17	–	25	–	–	–	27
C	–	25	–	26	–	–	–
D	–	–	26	–	15	–	28
E	–	–	–	15	–	18	11
F	–	–	–	–	18	–	13
G	28	27	–	28	11	13	–

(a) Choose any point, say A, and delete its row. Look down its column and select the smallest number. Note the new node to join A and draw the arc corresponding to the number selected.

(b) Delete the row of the node just chosen and look in the columns of the nodes selected so far for the smallest number. Choose this number and use the associated arc to connect a new node to the tree.

(c) Repeat (b) until 6 arcs have been chosen. (i.e. one less than the number of nodes.)

This process gives

	A	B	C	D	E	F	G
	↓						
B	(17)	–	25	–	–	–	27
C	–	25	–	26	–	–	–
D	–	–	26	–	15	–	28
E	–	–	–	15	–	18	11
F	–	–	–	–	18	–	13
G	28	27	–	28	11	13	–

A●

	A	B	C	D	E	F	G
	↓	↓					
C	–	(25)	–	26	–	–	–
D	–	–	26	–	15	–	28
E	–	–	–	15	–	18	11
F	–	–	–	–	18	–	13
G	28	27	–	28	11	13	–

A●————●B

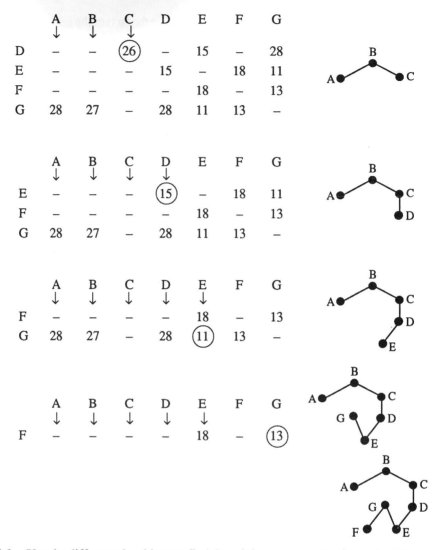

	A ↓	B ↓	C ↓	D	E	F	G
D	–	–	(26)	–	15	–	28
E	–	–	–	15	–	18	11
F	–	–	–	–	18	–	13
G	28	27	–	28	11	13	–

	A ↓	B ↓	C ↓	D ↓	E	F	G
E	–	–	–	(15)	–	18	11
F	–	–	–	–	18	–	13
G	28	27	–	28	11	13	–

	A ↓	B ↓	C ↓	D ↓	E ↓	F	G
F	–	–	–	–	18	–	13
G	28	27	–	28	(11)	13	–

	A ↓	B ↓	C ↓	D ↓	E ↓	F	G
F	–	–	–	–	18	–	(13)

Exercise 3.1 Use the different algorithms to find the minimum connector for each of the networks in Fig. 1.12 and give the length of each.

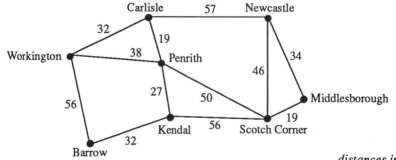

(a)

distances in miles

Fig. 1.12

(b)

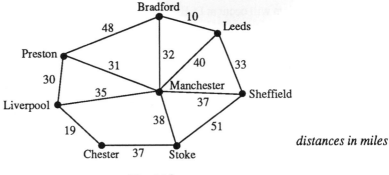

Fig. 1.13

(c)   Use the network in Fig. 1.10.

1.4   CHINESE POSTMAN PROBLEM

The **Chinese postman problem** involves finding a journey of minimum length that covers every arc. When finding the minimum journey you should avoid, if possible, traversing an arc more than once.

Networks can be put into three categories to describe their traversability. Any network has:

(a)   **no odd nodes**, so it is traversable starting and ending at the same point; the traversable path is called **Eulerian** (see section 1.6));

(b)   **two odd nodes**, in which case it is traversable starting and finishing at different points (**semi-Eulerian**);

(c)   **more than two odd nodes**, making it not traversable (**non-Eulerian**).

Can the postal worker leave the post office at P and walk along each of the roads in the village exactly once to deliver letters at each of the nine houses denoted by dots, finishing back at P?

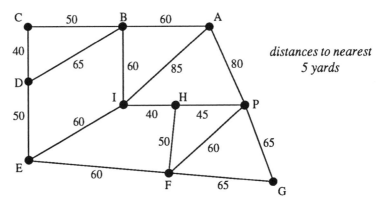

Fig. 1.14

The orders of the nodes are A3, B4, C2, D3, E3, F4, G2, H3, I4 and P4, so the network is not traversable. Leaving A for the second time will mean travelling along a road already walked and a similar problem will occur at D, E and H. Routes involving these four nodes will need to be repeated – but how can this be planned to minimise the distances travelled?

Repeating the routes $AD + EH = 125 + 100 = 225$ yards

$AE + DH = 145 + 150 = 295$ yards

$AH + DE = 125 + 50 = 175$ yards

so repeating the routes AH and DE involves the least extra walking. The total distance to be covered is the sum of all the lengths in the network plus 175 yards, a total of 1110 yards. One possible route would be P A I H P G F **H I A** B C D E I B **D E** F P with the repetitions shown in underlined bold type.

Another way of thinking about the problem is to ask, "Which arcs should I add to the network to make it Eulerian with the smallest possible total length, if the only new ones allowed are copies of those that already exist?".

The following network has six odd nodes, so you would need to consider pairings such as AB, CD, EF as these involve all the problem points. How many such arrangements are there?

In fact there are fifteen to consider, so looking at them all would be quite time consuming.

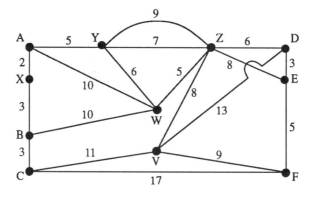

Fig. 1.15

With so many possibilities it is sensible to try to eliminate some combinations. In the case of the diagram in Fig. 1.15, it is not logical to consider any arrangement that links each node on the left with one on the right, so you can dismiss six cases, AD BE CF, AD BF CE, etc., which constitutes a considerable help.

Exercise 4.1   A railway system must be checked for fallen trees after bad weather. How should the inspector's route be planned if it is to start and finish at the depot, D?

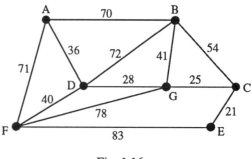

distances in kilometres

Fig. 1.16

Exercise 4.2   A motor manufacturer plans to paint its 'MN' logo on vehicles, using a robot. The machine's life can be extended by turning it on and off as infrequently as possible. The whole logo will be painted starting and finishing at A without any breaks. What is the minimum length the machine can paint?

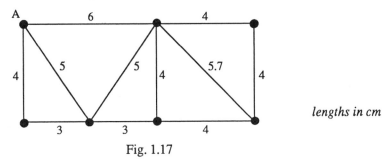

lengths in cm

Fig. 1.17

Exercise 4.3   Some holidaymakers want to see as much as possible of the island on which they are staying. They plan to use the bus service and want to ride all the routes possible, starting and ending at S. How many miles will they travel on their tour?

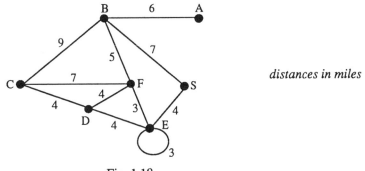

distances in miles

Fig. 1.18

1.5   TRAVELLING SALESMAN PROBLEM

The travelling salesman problem requires finding a route of minimum length which passes through every node. The optimal route may, in fact, pass through some nodes more than once. Many attempts have been made to find an algorithm to solve this problem, but without success.

Computers can assess all the **Hamiltonian cycles**, that is, circuits that go through each node exactly once, but there is no guarantee that the solution is of this form. Your work on this problem must be confined to

(a) finding an upper bound to the solution;

(b) finding a lower bound to the solution;

(c) finding the best answer you can without being sure that it is optimal.

Example

Find the shortest route starting and ending at Chepstow, which passes through each town at least once, using only the arcs given.

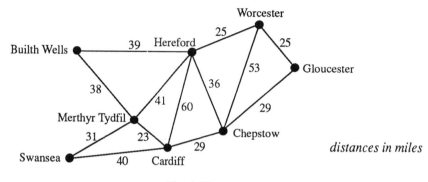

distances in miles

Fig. 1.19

Solution

(a) There are two methods for finding the **upper bound** or limit.

    (i) Start with the minimum spanning tree and double its length.

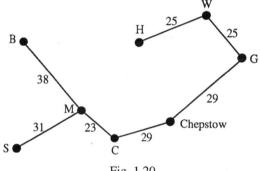

Fig. 1.20

The route Chepstow–G–W–H–W–G–Chepstow–C–M–B–M–S–M–C–Chepstow goes over every arc of the spanning tree exactly twice, giving a length of $(25 + 25 + 29 + 29 + 23 + 31 + 38) \times 2 = 400$ miles, and is not likely to be optimal. Therefore the length of the shortest route ≤ 400 miles, which is an upper bound, or limit, for the problem.

   This can be improved upon by going directly from S to C instead of via M, which saves $31 + 23 - 40 = 14$ miles, so the length of the shortest route ≤ 386 miles.

   Further improvements can reduce the upper bound; for example, going from H to C directly instead of via W, G and Chepstow, saving another 48 miles.

(ii) Find any route that passes through each node at least once.

The route Chepstow–G–W–H–B–M–S–C–Chepstow passes through each town and has a length $29 + 25 + 25 + 39 + 38 + 31 + 40 + 29 = 256$ miles. Either this is the optimal solution or it is possible to find a better one, so the length of shortest route ≤ 256 miles.

(b) Erase any node (for example, Hereford) and any associated arcs from the network and
     find the minimum connector for the reduced diagram.

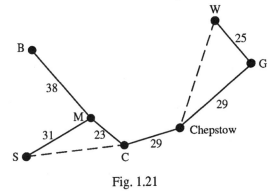

Fig. 1.21

Find the length of the minimum connector and add the lengths of the two shortest arcs from
H. This gives a lower bound of $(38+31+23+29+29+25)+25+36 = 236$ miles.
By deleting different nodes a selection of lower bounds can be found. For example,
deleting Chepstow gives the network in Fig. 1.22 and a minimum connector of length
$25+25+39+38+31+23 = 181$ miles; adding the two shortest deleted arcs gives
$181+29+29 = 239$ miles.

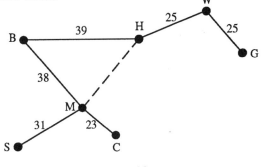

Fig. 1.22

Since the length of the shortest route ≥ 236 miles and the length of shortest route
≥ 239 miles, the second inequality is chosen and the first is discarded, as it gives less
information.
     If there had been no need to visit Swansea, the network would be as shown in Fig. 1.23.

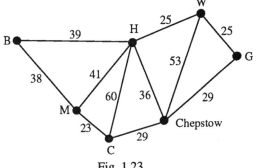

Fig. 1.23

Deleting Chepstow and finding the minimum spanning tree shows that it is, in fact, a chain. Adding the two shortest arcs from Chepstow joins the town to the ends of the chain and gives an optimal route length of 208 miles.

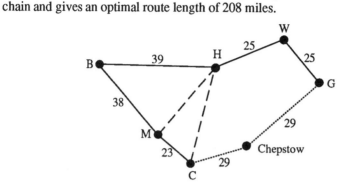

Fig. 1.24

The minimum connector will not be a chain, or the shortest deleted arcs will not go to its ends. Some arcs will need to be traversed more than once and so the calculated length will be less than that actually required, i.e. a lower bound. The shortest route for the original problem satisfies

$$239 \leq \text{shortest route} \leq 256 \text{ miles.}$$

Note: A map such as that shown in Fig. 1.25 can be redrawn as in Fig. 1.26, so that all arcs go from one node to another node.

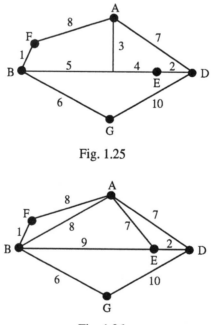

Fig. 1.25

Fig. 1.26

Exercise 5.1 By deleting the towns one at a time in turn from this network, consider the sets of five nodes that remain each time and calculate a lower bound for the travelling salesman problem in each case.

Choose the one that gives the most useful value as a lower bound. Is this an optimal value (i.e. was it formed by arcs from the deleted node joining the ends of a chain to form a circuit)?

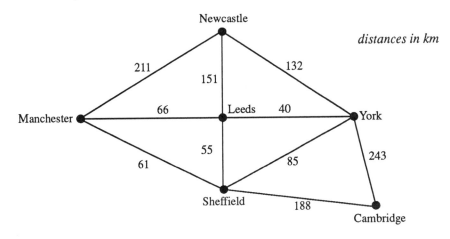

Fig. 1.27

Exercise 5.2 Find the minimum spanning tree for this network and use it to give an upper bound for the travelling salesman problem. Make any improvements to this bound that you can.

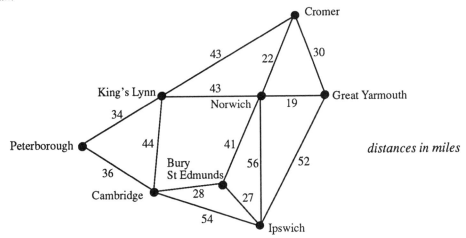

Fig. 1.28

Exercise 5.3 Ann works for Wellies the Chemists and has to visit branches at all the towns shown, leaving and returning to her office in Exeter. What is the least distance she needs to travel? Would it help if her office were moved to Okehampton to be more central?

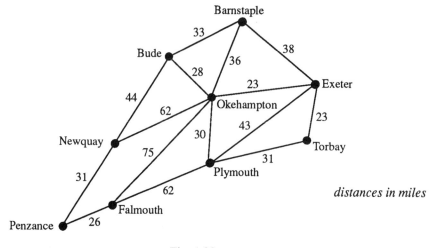

Fig. 1.29

Exercise 5.4 Plan a route for A. Merican who is staying in London and wants to tour the South East of England visiting the towns shown, and then return to London. How many miles need be travelled?

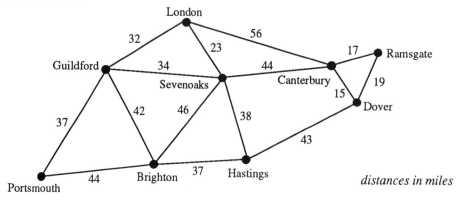

Fig. 1.30

Exercise 5.5 After a very cold night, a railway-track inspector needs to check all the lines in his part of the system, as shown in the network in Fig. 1.31.

(a) If he starts and finishes at A, how many miles will he cover if he uses the shortest possible route?

(b) If he sets out to check only the points at junctions, how far need he travel?

(c) If he is called to check the line at D as well as the points, will this increase the distance
he covers?

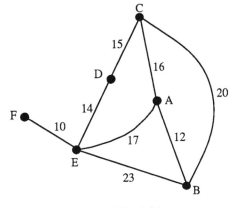

Fig. 1.31

Exercise 5.6   On arrival at a bird sanctuary, visitors leave the parking area, A, and use the
paths as shown to reach the hides B, C, D and E. The times taken to walk along the different
paths, in minutes, are given beside each path. Redraw the network so every node is connected
to every other one directly.

Find a route which visits every hide in the least time. Is the optimal solution unique?

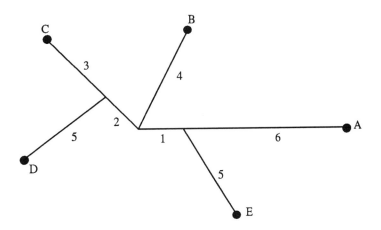

Fig. 1.32

1.6   NOTES

Leonard Euler was a Swiss mathematician of the eighteenth century, remembered for his work in mechanics, calculus and geometry. The result for networks drawn on a plane that

'number of nodes plus the number of regions is equal to the number of arcs plus 2'

is Euler's formula (where the outside region is included in the total number).

If a network is Eulerian, or has been made so by the addition of arcs, the problem remains of how to find a path containing each line exactly once. This may be solved by using **Fleury's algorithm**, which says

(a)   start at any node and move along the arcs according to the rules;

(b)   (i) erase the arcs as they are used;
      (ii) if using an arc disconnects the network, only do so when there is no other choice;

(c)   repeat (b) until all arcs have been erased.

William Hamilton was a nineteenth century Irish mathematician, physicist and astronomer who made major contributions to mechanics and calculus.

A network is said to be **Hamiltonian** if it contains a circuit that passes through each node exactly once. While it is easy to see if a network is Eulerian or not by checking whether it has no odd nodes, no such simple test has been found to determine if a network is Hamiltonian. Theories about 'Hamiltonian-ness' do exist, though. **Dirac's** theorem, for example, states that if a network has at least three nodes and the order of every node is no less than half the number of nodes, then it is Hamiltonian. (So a network with six nodes will certainly be Hamiltonian if the order of every node ≥ 3.)

2

Recursion

2.1 DEFINITION

In general English the word **recursion** means **'the act of returning'**. In mathematics it may also be used in going forwards, as the term is applied to the step by which each element of a sequence is generated from its precedent. This step may then be taken either backwards or forwards, so the word gains a wider meaning.

2.2 INVESTIGATION

In a board game, a counter is moved two squares forward if the die shows a five or a six otherwise it moves one place forward. Let $P(n)$ be the probability that the counter lands on a square numbered n. It starts on the square labelled zero. How can $P(4)$ be found? Write down the values of $P(0)$ and $P(1)$. Explain why $P(4) = \dfrac{2}{3}P(3) + \dfrac{1}{3}P(2)$.

This means that if the values of $P(3)$ and $P(2)$ are known, then $P(4)$ can be found. So the problem should be easier as it is now possible to find the values of $P(2)$ and $P(3)$. Write down similar expressions for $P(3)$ and $P(2)$. Now use these to evaluate $P(2), P(3)$ and $P(4)$. What do you think the value of $P(100)$ would be?

How would the values be changed if the probability of moving two places was $\frac{1}{4}$? (Chapter 12 on *Recurrence Relations* shows further similar work.)

2.3   DIVISIBILITY

Example

A number is divisible by 13 if the sum of four times the units digit and the number formed by the remaining digits, is divisible by 13.

Is $12\,382\,247\,310\,082$ divisible by 13?

Solution

The problem can be reduced to,

    is $1238\,224\,731\,008 + 4 \times 2 = 1\,238\,224\,731\,016$, divisible by 13?

Further reductions lead to

$$123\,822\,473\,101 + 4 \times 6 = 123\,822\,473\,125$$
$$12\,382\,247\,312 + 4 \times 5 = 12\,382\,247\,332$$
$$1\,238\,224\,733 + 4 \times 2 = 1\,238\,224\,741$$

This ten-digit number can be reduced further and, as it fits on any calculator screen, you can easily find that

$$1\,238\,224\,741 \div 13 = 95\,248\,057$$

The answer to the original question is 'yes'!

2.4   HIGHEST COMMON FACTORS   (EUCLID'S ALGORITHM)

A method of finding the **highest common factor** (HCF) of two numbers is to use the highest common factor of the smaller, and the remainder when the larger is divided by the smaller.

Example

Find the highest common factor (HCF) of 19302 and 2508.

Solution

The division $19\,302 \div 2508 = 7$ remainder 1746, so the highest common factor of 19302 and 2508 is the highest common factor of 2508 and 1746. Since the numbers are smaller, the problem should be easier to solve.

        $2508 \div 1746 = 1$ remainder 762

$\Rightarrow$ HCF of 2508 and 1746 $=$ HCF of 1746 and 762.

        $1746 \div 762 = 2$ remainder 222

$\Rightarrow$ HCF of 1746 and 762 $=$ HCF of 762 and 222

        $762 \div 222 = 3$ remainder 96 $\Rightarrow 222, 96$

        $222 \div 96 = 2$ remainder 30 $\Rightarrow 96, 30$

        $96 \div 30 = 3$ remainder 6 $\Rightarrow 30, 6$

        $30 \div 6 = 5$ remainder 0 $\Rightarrow 6, 0$

$\Rightarrow$ HCF $= 6$

Recursive procedures will be met in many areas of discrete maths, notably in dynamic programming, iterative methods and recurrence relations.

Exercise 4.1   A number is divisible by seven if the difference between twice the unit digit and the number formed by the remaining digits, is divisible by seven. Are these numbers divisible by 7?

(a)   397 639 828 194

(b)   16 555 398 868 681

Exercise 4.2   A number is divisible by eleven if the difference between the sums of the digits in the odd and even places is zero or a multiple of eleven. Are these numbers divisible by 11?

(a)   724 793 536 016

(b)   25 698 741 392

Exercise 4.3   Is 208 146 725 361 a perfect square?
(Hint: divide by the squares of the primes, each one repeatedly, until a square or otherwise is recognised.)

Exercise 4.4   The labels on the arcs are distances in miles. Find the shortest route from S to T by finding the shortest routes to T from A and B. (See Chapter 4 on *Dynamic Programming* for further examples of this topic.)

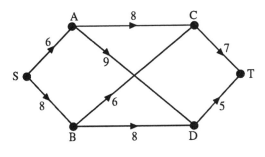

Fig. 2.1

Exercise 4.5   A network is described as **planar** if it can be drawn with no arcs crossing each other. Euler's relation for connected planar networks states that $N + R = A + 2$ where $N =$ number of nodes, $R =$ number of regions and $A =$ number of arcs.

If a network consists of just one node and no arcs as in Fig. 2.2, then $N = 1$, $A = 0$ and $R = 1$ (the surrounding region is always included in the count), and $1 + 1 = 0 + 2$.

Draw any connected planar network on seven nodes, or use the one given in Fig. 2.2, as follows.

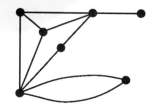

Fig 2.2

Remove the arcs one at a time, keeping the network connected as long as possible. When the network has to be broken, break it so that single nodes are separated from the rest and delete them. Check that, when each arc is removed, the number of either nodes or regions is reduced by one, i.e. when A is reduced by 1, so is $N + R$. Stop when ● has been reached.

You should now see that Euler's relation always holds for connected planar networks.

Exercise 4.6   The highest common factor of two numbers is the same as the highest common factor of the smaller of the two and the difference between the two numbers. Use this idea to find the HFC of 1323 and 1155.

2.5 NOTES

Euclid's algorithm also applies to polynomials.

Euler's relation also applies to surfaces and to networks which are non-planar in the form
$$N + R = A + 2 - 2x - y$$
where x is the number of **handles** and y is the number of **cross-caps**. A **handle** is just what you think it is, i.e. a hole in a surface, so a doughnut has one handle and a figure of eight has two. A **cross-cap** is more difficult to visualize, but can be considered as the result of pulling part of a sphere through a hole in its surface.

The rules quoted for divisibility by 7 and 13 can be proved as follows. ($a \mid b$ means that a divides exactly into b with no remainder.)

$$7 \mid 10x + y \Leftrightarrow 7 \mid 10x + y - 21y$$
$$\Leftrightarrow 7 \mid 10x - 20y$$
$$\Leftrightarrow 7 \mid x - 2y$$

$$13 \mid 10x + y \Leftrightarrow 13 \mid 10x + y + 39y$$
$$\Leftrightarrow 13 \mid 10x + 40y$$
$$\Leftrightarrow 13 \mid x + 4y$$

Can you devise a rule for testing the divisibility of numbers by higher primes?

3

Shortest route

3.1 INVESTIGATIONS
Investigation 1
Find the shortest route from Peterborough to Liverpool.

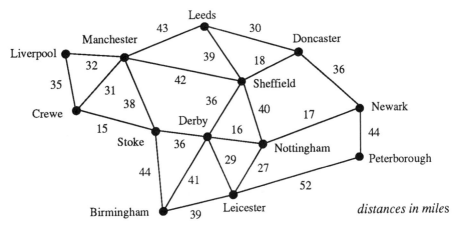

Fig. 3.1

Investigation 2

The network in Fig. 3.2 shows a system of roads between towns and the approximate travelling times between them in hours. Find the quickest route from S to T.

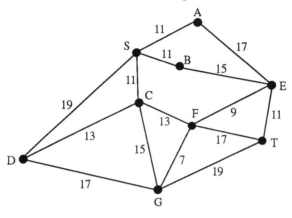

Fig. 3.2

3.2   DIJKSTRA'S ALGORITHM

Most methods of tackling shortest route problems produce not only the route but also its length. Dijkstra's method works its way across a network from the start node to the finish node through intermediate points in order of increasing distance.

Step (a)   Label the start node 'zero'.

Step (b)   Calculate working values for each node directly connected to the one just labelled, using working value = minimum of: latest label + distance from labelled node.

Step (c)   Select the unlabelled node with the lowest working value and label it with that value.

Step (d)   Repeat Steps (b) and (c) until the finish node has been labelled, then stop.

Step (e)   To find the shortest route, trace back from the destination node to the start along those arcs which satisfy
length of arc = difference of labels.

Essentially this method begins by asking, "Which node is closest to the start?". There can be no quicker way to reach this point, so now ask, "Which node is closest to these two". The process then works its way across the network.

There is no reason why every node in the network should be labelled. Any left unlabelled are further from the start node than is the destination. It is quite possible that more than one node will be a candidate for labelling. Take any with the lowest working value. The trace-back may well reveal more than one optimal route from finish to start. Optimal routes may have some arcs in common or be completely disjoint.

Solution of Investigation 2

Find the shortest time and associated route for a journey from S to T.

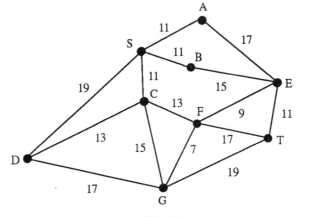

*all times to the
nearest hour*

Fig. 3.3

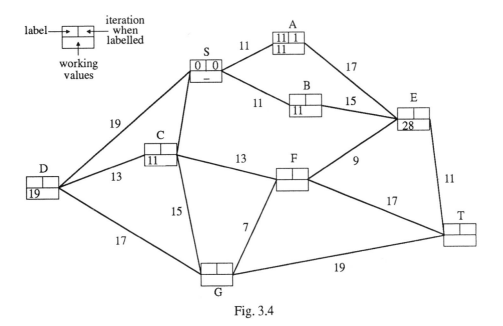

Fig. 3.4

The start node has been labelled with zero. Any of A, B, C may have been assigned the label 11. A has been chosen and further working values calculated from there. For example, E, the only node connected to A, is given the working value $11+17 = 28$. Continue the example on a diagram of your own until T is labelled 37.

Trace-back gives T - E - B - S as label T – label E = 'length' of E T, etc., so the shortest route is S - B - E - T, which takes 37 hours.

Exercise 2.1   Use Dijkstra's algorithm to find the shortest path from S to T.

(a)

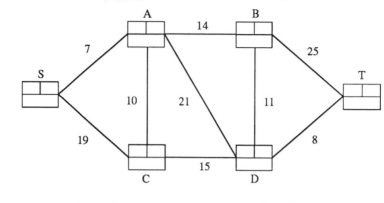

Trace-back: Length:

Fig. 3.5

(b)

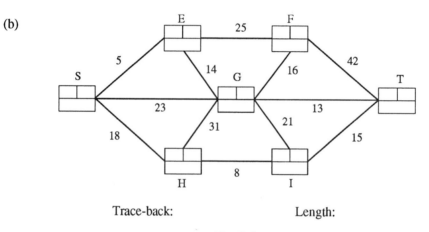

Trace-back: Length:

Fig. 3.6

Exercise 2.2   Find the shortest route from York to Carlisle which does not pass through Scotch Corner. What is the shortest route via Scotch Corner?

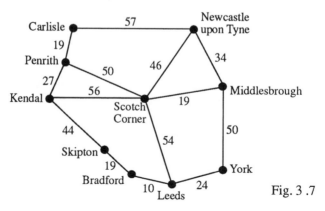

Fig. 3 .7

Exercise 2.3   Use Dijkstra's algorithm to solve Investigation 1.

3.3   DELAYS AT NODES
Example

A company employee frequently moves between two offices, S and T, in London. There are two car parks, A and B, the first of which is three minutes walk from S and the second two minutes from S. On average, it takes one minute to drive out of car park A and three minutes to drive out of car park B. Car parks F and H are each three minutes from T; car park G is four minutes from T.

The numbers in Fig. 3.8 give times between points on the employee's possible routes. From experience, she know that she can expect a two minute delay at H.

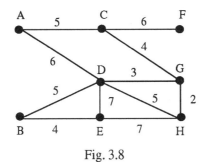

Fig. 3.8

(a)   Redraw the network showing all the information given, and use Dijkstra's algorithm to find the quickest route from S to T.

(b)   If there is a three minute delay at D, will this affect the minimum possible travelling time?

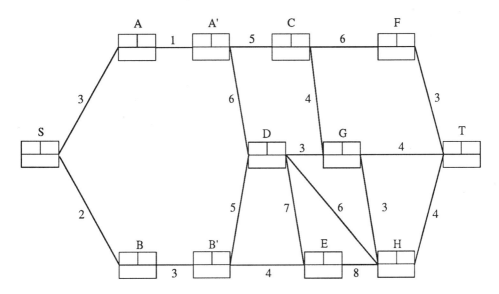

Fig. 3.9

Solution

Two ways of adapting the network to take account of the delays described are shown in the network in Fig. 3.9. The two minute delay at H has been incorporated into the figures by adding one to each arc that is incident with it. Since any route involving H will use two such arcs, two minutes will have been added as required. Similarly, a ten minute delay would be included by adding five to each arc that meets the node.

The times to leave A and B could have been dealt with in the same way by adding half a minute to each arc meeting A and one and a half minutes to each incident with B. Instead, new arcs and nodes have been introduced to show the delays there.

This could not have been done at H as a route might enter along arc (a) and leave via (b) without experiencing the required delay.

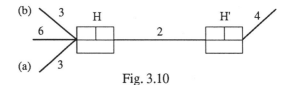

Fig. 3.10

3.4 CHINESE POSTMAN PROBLEM
Example

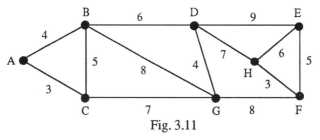

Fig. 3.11

Orders of nodes: A2, B4, C3, D4, E3, F3, G4, H3.

Find the shortest route starting and ending at A that traverses all the arcs.

Solution

This network has four odd nodes, C, E, F and H, and since three are on the right of the diagram and one on the left, the question is what are the shortest distances from C to each of E, F and H. This is a case where Dijkstra's algorithm starting at C would be appropriate.

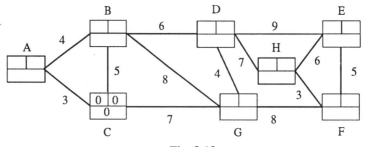

Fig. 3.12

Now it is possible to consider the pairings

$$\left.\begin{matrix} CE \\ FH \quad 3 \end{matrix}\right\} \qquad \left.\begin{matrix} CF \\ EH \quad 6 \end{matrix}\right\} \qquad \left.\begin{matrix} CH \\ EF \quad 5 \end{matrix}\right\}$$

and find a path starting and ending at C.

When the shortest distance between all pairs of points is required, Dijkstra's algorithm may be used with each node (except one) as the start but it is more efficient to use Floyd's algorithm (see section 3.6 *Notes*, for details of this algorithm), which can also be used in situations like this.

3.5 ARCS WITH NEGATIVE VALUES
Example

Sassan wants to get from city S where he is a student, to town T where his family live.

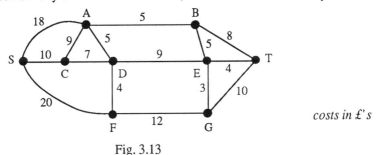

costs in £'s

Fig. 3.13

He is short of money and must find the cheapest route home. The costs of journeys between the towns are shown on the arcs.

He knows that if he calls on his uncle Soheil at G then he will get £10 from him. How should he plan his journey?

Solution

Taking £5 from each arc incident with G and applying Dijkstra's algorithm gives the network in Fig. 3.14.

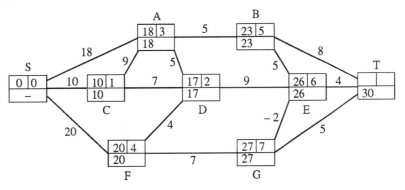

Fig. 3.14

At this stage, G has just been labelled 27, which provides a route of length 25 to E, which has already been labelled 26. The method has failed to produce the nodes in ascending distances from S, and therefore breaks down in this case. Adding three to each arc will avoid negative numbers but a route using four arcs will have twelve added to its length, while another of five will have its total increased by fifteen, so this idea fails. Another method such as dynamic programming (see Chapter 4) or Floyd's algorithm, is needed.

Exercise 5.1   Complete the example in section 3.3.

Exercise 5.2   Complete the example in section 3.4.

Exercise 5.3
(a) In Fig. 3.1 the numbers represent times in minutes. Find the quickest route from P to Q.

(b) Which route is quickest if there is a 6 minute delay at A?

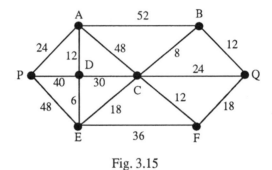

Fig. 3.15

Exercise 5.4   Fig. 3.16 shows the time, in minutes, for bus journeys around a town. A change of bus is required at each vertex. If each change involves a two minute wait, find the quickest route from S to T.

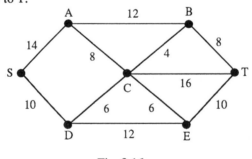

Fig. 3.16

3.6   NOTES

Dijkstra's algorithm may be adapted to find the longest path through a network for situations where the numbers on arcs represent, for example, profits. At the start, label the initial node zero and at each stage consider those nodes that can be reached only from nodes already labelled. Choose the one with the **highest** working value to assign a label.

It should be noted that subtracting the value on each arc from the largest in an attempt to create a minimization problem, will not give a correct solution.

Floyd's algorithm

Example

Describe the network in Fig. 3.17 by means of a matrix giving direct distances between nodes where such arcs exist, and ∞ where they do not.

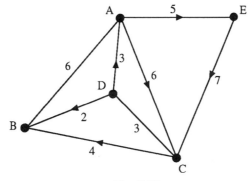

Fig. 3.17

Solution

Use A's row and columns. Add each entry in the row to corresponding entries in the column below and actually enter the number in the array if it is lower than the one already there.

TO

FROM

	A	B	C	D	E
A	0	6	6	∞	5
B	6	0	∞	∞	∞
C	∞	4	0	3	∞
D	3	2	3	0	∞
E	∞	∞	7	∞	0

↓

	A	B	C	D	E
→ A	0	6	6	∞	5
B	6	0	(12)	∞	(11)
C	∞	4	0	3	∞
D	3	2	3	0	(8)
E	∞	∞	7	∞	0

In this case there are three changes, which have been shown ringed. E and C's rows and D's column cannot change due to the ∞ in the first cell.

Repeat the process using B's row and column. This time there will be no change in D's column or E's row due to the ∞ marked with an asterisk (*).

$\downarrow$

	A	B	C	D	E
A	0	6	6	∞	5
→B	6	0	12	∞*	11
C	(10)	4	0	3	(15)
D	3	2	3	0	8
E	∞	∞	7	∞	0

$\downarrow$

	A	B	C	D	E
A	0	6	6	(9)	5
B	6	0	12	(15)	11
→C	10	4	0	3	15
D	3	2	3	0	8
E	(17)	(11)	7	(10)	0

$\downarrow$

	A	B	C	D	E
A	0	6	6	9	5
B	6	0	12	15	11
C	(6)	4	0	3	11
D	3	2	3	0	8
E	(13)	11	7	10	0

$\downarrow$

The final matrix gives the lengths of the shortest routes between any pairs of points.

	A	B	C	D	E
A	0	6	6	9	5
B	6	0	12	15	11
C	6	4	0	3	11
D	3	2	3	0	8
→E	13	11	7	10	0

To find the route of length 13 from E to A, find where the 13 first arose. It appeared by adding (E to D)10 to (D to A)3 when D's row and column were used. Now look back to see how they were formed. D to A has been unchanged throughout, but E to D became 10 by adding (E to C)7 to (C to D)3. These are unchanged since the start, so the route is E→C→D→A.

4

Dynamic programming

4.1 INVESTIGATIONS
Investigation 1
Find the shortest distance from X to Y.

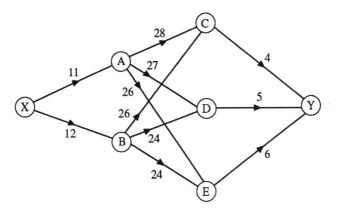

Fig. 4.1

Investigation 2

Find the shortest distance from S to T.

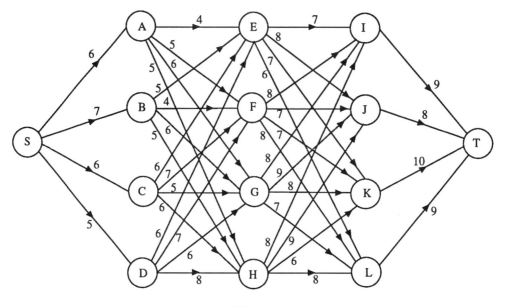

Fig. 4.2

For simple problems it is easy to find the shortest distance but for more complex networks you need an algorithm.

4.2   THE METHOD OF DYNAMIC PROGRAMMING

The basic idea behind dynamic programming is **Bellman's principle,** which states that the optimal solution to an n step process can be derived from the optimal solution to an $(n-1)$ step process with the optimal choice for the first step. Essentially, a problem with n steps may be reduced to one with $n-1$ steps, which may itself be reduced to $n-2$ steps, etc.

Example

There are 15 matches on the table in front of you. You start by picking up 1, 2 or 3 of them. Your opponent then does the same and so on until the last match is picked up by the 'loser'.
     What should be your strategy?

Solution

Leaving one match for your opponent to pick up will give you a win. Working backwards from here you can see that leaving five matches will also guarantee success as, however many your opponent removes, x, you can take a number, y, so that one and only one will be left. Now the problem is smaller. With 15 matches on the table, how can you leave your opponent with 5? (This is, of course, like starting with 11 and aiming to leave just one.) Repeating the argument shows that leaving 9 will result in a win and, similarly, leaving 13 will also win, so the correct first move is to take two.

This shows how a problem may be reduced in size and also illustrates the value of working backwards from the desired outcome.

Example: Investigation 2

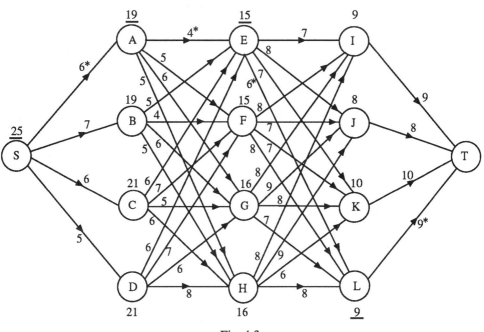

Fig. 4.3

How can T be reached? I is 9 units from T, J is 8 units from T, K is 10 units from T and L is 9, so we can see how to get to T from these nodes. What is the best way to reach T from E, F, G, and H? E is 15 units from T by going via L, and other routes are no shorter.

The shortest routes from F, G and H are 15, 16 and 16, respectively. Using the numbers now assigned to these nodes we can calculate the distances of A, B, C and D from T.

Routes from A to T go via E, F, G or H and have lengths $4+15$, $5+15$, $6+16$ and $5+16$ units, so A is 19 units from T.

Similarly B is 19 units and C and D are each 21 units from T. These values can now be used to find the best route from S.

The possible lengths are $6+19$, $7+19$, $6+21$ and $5+21$ units, so the shortest is 25, via A. The 19 units at A came from $4+15$ via E, where the 15 was the result of $6+9$ via L, so the optimal route is S A E L T. In Fig. 4.3, the arcs on this route are marked with asterisks (*) and the nodes have their assigned values underlined.

There may be more than one optimal route.

The advantage of dynamic programming over simply trying all the possibilities is that it requires far fewer calculations. For example, to try the $4 \times 4 \times 4 = 64$ different paths through the network given here would involve $64 \times 3 = 192$ additions, since each route needs 3 additions; e.g. S A E I T: $6+4+7+9 = 26$.

The method just used required four additions at each of E,F,G,H,A,B,C,D and S, giving a total of $9 \times 4 = 36$. The actual number of additions has been reduced to less than twenty per cent of that required before!

Dynamic programming may just as easily be used to find the **largest** route through a network in cases where the numbers associated with the arcs represent, for instance, profits.

Exercise 2.1   (a) Find the shortest route from S to T.

(b) Find the longest route from S to T.

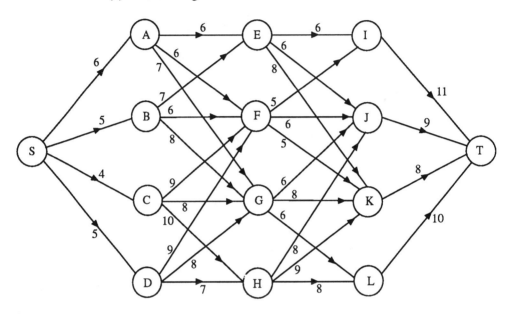

Fig. 4.4

Exercise 2.2   (a) Find the shortest route from S to T.

(b) Find the longest route from S to T.

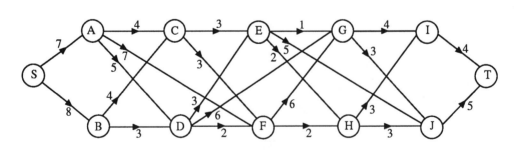

Fig. 4.5

Exercise 2.3   In the network in Fig. 4.6, you have to travel from town A to town E and, in doing so, have to pass through intermediate towns. What is the minimum cost route, given that you know the route costs between adjacent towns?

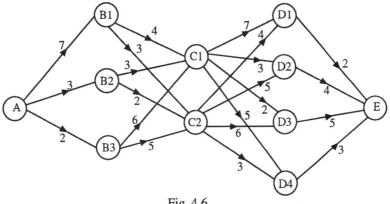

Fig. 4.6

4.3   APPLICATIONS OF DYNAMIC PROGRAMMING

Dynamic programming is a technique that can be used to solve many **optimisation** problems. It may be used when the network or problem can be split into **stages** and there is a direction and cost/return on each arc which is identified by an **action**.

If you can identify the different **states** involved, you have every chance of being able to draw a network to describe the problem. A typical network formulation of a problem is shown in Fig. 4.7.

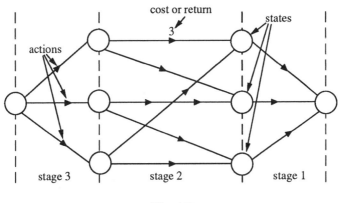

Fig. 4.7

A tableau may be drawn up showing all the calculations involved in solving the problem, including those which produce non-optimal values which are not assigned to nodes. With practice, a tableau may be drawn up and a problem solved **without** the need to visualise the problem by means of a diagram.

Example

After moving into a new house, a family decides to build three walls around the garden to make it safer for the children. Work done on the later walls may be helped by expertise gained on the earlier ones, but there may be other factors such as the state of adjacent flower beds and the amount of local traffic.

In the next four weeks the husband will have to spend one week at work, so they plan to put up one wall in each of the other three weeks.

	Week 1	Week 2	Week 3	Week 4
Wall A	15	15	17	15
Wall B	8	13	11	11
Wall C	15	17	15	14

(a) Identify appropriate stages, states, actions and returns for a dynamic programming formulation of the problem.

(b) Draw a network to represent the project using the figures from the table above. Show clearly what the nodes and arcs of the network represent.

(c) Set up a tableau to solve this problem using dynamic programming. Solve the problem and give the best schedule for completing the project.

(d) What network problem corresponds to the dynamic programming problem?

Method

(a) states (nodes) ≡ walls built (either 0 for none or A, B, C or a combination of these)

  stages ≡ weeks (e.g. 1, 2, 3 or 4)

  actions (arcs) ≡ wall to build that week

  returns ≡ number of hours required for action

(b)

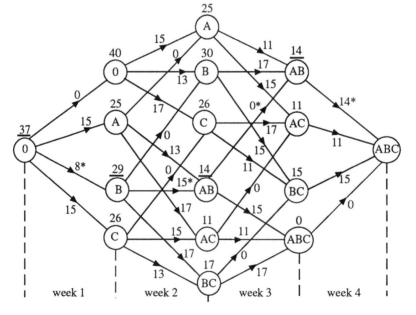

Fig. 4.8

So, for week 1, the four possible actions give

no wall built,   wall A built,   wall B built,   wall C built,

each corresponding to an arc from the initial state. The hours needed for each action are given on the corresponding arc. From each state, after week 1, the possible actions are again constructed and this continues until the final state, ABC, at the end of week 4.

(c)   The information from the network can be analysed in a table.

Stage (week)	State (walls built)	Action (to build)	Time
1	ABC	0	0
(week 4)	AB	C	$14 \leftarrow$
	AC	B	11
	BC	A	15
2	AB	0	$0 + 14 = \underline{14} \leftarrow$
		C	$15 + 0 = 15$
	AC	0	$0 + 11 = \underline{11}$
		B	$11 + 0 = \underline{11}$
	BC	0	$0 + 15 = \underline{15}$
		A	$17 + 0 = 17$
	A	B	$11 + 14 = \underline{25}$
		C	$15 + 11 = 26$
	B	A	$17 + 14 = 31$
		C	$15 + 15 = \underline{30}$
	C	A	$17 + 11 = 28$
		B	$11 + 15 = \underline{26}$
3	A	0	$0 + 25 = \underline{25}$
		B	$13 + 14 = 27$
		C	$17 + 11 = 28$
	B	0	$0 + 30 = 30$
		A	$15 + 14 = \underline{29} \leftarrow$
		C	$17 + 15 = 32$
	C	0	$0 + 26 = \underline{26}$
		A	$15 + 11 = \underline{26}$
		B	$13 + 15 = 28$
	0	A	$15 + 25 = \underline{40}$
		B	$13 + 30 = 43$
		C	$17 + 26 = 43$
4	0	0	$0 + 40 = 40$
		A	$15 + 25 = 40$
		B	$8 + 29 = \underline{37} \leftarrow$
		C	$15 + 26 = 41$

To find the best schedule, you start at week 4, i.e. state 1. The minimum time is 14, and this corresponds to action C in week 4. This means that you have come from state AB (starred on diagram and arrowed on table). During week 3, the optimum policy is to build nothing, in week 2 to build A and in week 1 to build B. This is summarised below.

Best schedule:	week	1	2	3	4
	wall	B	A	0	C

Total time = 37 hours.

(d) The dynamic programming problem is equivalent to finding the shortest route through the network.

Example (a stock control problem)

A manufacturer knows that for the next 4 months the demand for a product is as follows:

month	1	2	3	4
demand	1	3	2	4

With each production run there is a setup cost of £600, as well as a production cost of £200 per unit made. At the end of each month there is a holding cost of £100 per unit still in stock, having met the orders for that month. Production capacity limits production to, at most, 5 units per month, and warehouse restrictions mean that no more than 4 units can be kept in stock. How should production be planned to meet the demands at minimum cost if there are no units in stock at the outset?

Method

(a) Identify the stages.

(b) Identify the states and actions.

(c) Decide whether or not to draw a network to describe the problem.

(d) Calculate the costs and use dynamic programming to find the optimal production schedule.

Solution

(a) stages   ≡ months (denoted by S, A, B, C in Fig. 4.9)

(b) states   ≡ number in stock at the start of the stage (shown as a number in the node)

    actions ≡ number to build that month

(c) see Fig. 4.9

(d) calculation follows Fig. 4.9.

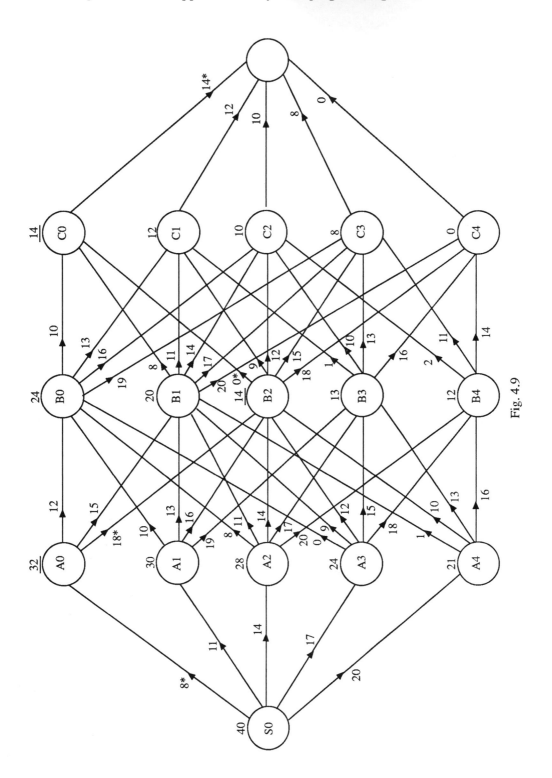

Fig. 4.9

(d)   *Costs in 100s of £s*

Stage	Start state	Action	Production and holding cost	Total cost	End state
1	0	4	$14 + 0 = 14$	14	0
	1	3	$12 + 0 = 12$	12	0
	2	2	$10 + 0 = 10$	10	0
	3	1	$8 + 0 = 8$	8	0
	4	0	$0 + 0 = 0$	0	0
2	0	2	$10 + 0 = 10$	$10 + 14 = 24$ *	0
	0	3	$12 + 1 = 13$	$13 + 12 = 25$	1
	0	4	$14 + 2 = 16$	$16 + 10 = 26$	2
	0	5	$16 + 3 = 19$	$19 + 8 = 27$	3
	1	1	$8 + 0 = 8$	$8 + 14 = 22$	0
	1	2	$10 + 1 = 11$	$11 + 12 = 23$	1
	1	3	$12 + 2 = 14$	$14 + 10 = 24$	2
	1	4	$14 + 3 = 17$	$17 + 8 = 25$	3
	1	5	$16 + 4 = 20$	$20 + 0 = 20$*	4
	2	0	$0 + 0 = 0$	$0 + 14 = 14$ *	0
	2	1	$8 + 1 = 9$	$9 + 12 = 21$	1
	2	2	$10 + 2 = 12$	$12 + 10 = 22$	2
	2	3	$12 + 3 = 15$	$15 + 8 = 23$	3
	2	4	$14 + 4 = 18$	$18 + 0 = 18$	4
	3	0	$0 + 1 = 1$	$1 + 12 = 13$*	1
	3	1	$8 + 2 = 10$	$10 + 10 = 20$	2
	3	2	$10 + 3 = 13$	$13 + 8 = 21$	3
	3	3	$12 + 4 = 16$	$16 + 0 = 16$	4
	4	0	$0 + 2 = 2$	$2 + 10 = 12$*	2
	4	1	$8 + 3 = 11$	$11 + 8 = 19$	3
	4	2	$10 + 4 = 14$	$14 + 0 = 14$	4

Costs in 100s of £'s

Stage	Start state	Action	Production and holding cost	Total cost	End state
3	0	3	$12 + 0 = 12$	$12 + 24 = 36$	0
	0	4	$14 + 1 = 15$	$15 + 20 = 35$	1
	0	5	$16 + 2 = 18$	$18 + 14 = 32*$	2
	1	2	$10 + 0 = 10$	$10 + 24 = 34$	0
	1	3	$12 + 1 = 13$	$13 + 20 = 33$	1
	1	4	$14 + 2 = 16$	$16 + 14 = 30*$	2
	1	5	$16 + 3 = 19$	$19 + 13 = 32$	3
	2	1	$8 + 0 = 8$	$8 + 24 = 32$	0
	2	2	$10 + 1 = 11$	$11 + 20 = 31$	1
	2	3	$12 + 2 = 14$	$14 + 14 = 28*$	2
	2	4	$14 + 3 = 17$	$17 + 13 = 30$	3
	2	5	$16 + 4 = 20$	$20 + 12 = 32$	4
	3	0	$0 + 0 = 0$	$0 + 24 = 24*$	0
	3	1	$8 + 1 = 9$	$9 + 20 = 29$	1
	3	2	$10 + 2 = 12$	$12 + 14 = 26$	2
	3	3	$12 + 3 = 15$	$15 + 13 = 28$	3
	3	4	$14 + 4 = 18$	$18 + 12 = 30$	4
	4	0	$0 + 1 = 1$	$1 + 20 = 21*$	1
	4	1	$8 + 2 = 10$	$10 + 14 = 24$	2
	4	2	$10 + 3 = 13$	$13 + 13 = 26$	3
	4	3	$12 + 4 = 16$	$16 + 12 = 28$	4
4	0	1	$8 + 0 = 8$	$8 + 32 = 40*$	0
	0	2	$10 + 1 = 11$	$11 + 30 = 41$	1
	0	3	$12 + 2 = 14$	$14 + 28 = 42$	2
	0	4	$14 + 3 = 17$	$17 + 24 = 41$	3
	0	5	$16 + 4 = 20$	$20 + 21 = 41$	4

Optimal schedule :   1, 5, 0, 4 $\Rightarrow$ cost £4000

Exercise 3.1 A company makes wooden garden sheds of three types, A, B and C. There are orders for one shed of each type, and it is planned to make them in the order which will maximise profits. If shed B is made first there will be a profit of £65, but if shed B is made after shed A, then some left-over materials can be used and the profit will rise to £70. Profit details are given below.

				Already built			
Shed	None	A	B	C	AB	AC	BC
A	58	-	60	65	-	-	70
B	65	70	-	70	-	85	-
C	70	86	80	-	90	-	-

What is the optimal order of construction and hence the maximum profit?

Exercise 3.2 A maths student has four text books, P, Q, R and S, to work through in her holiday. The books are on related topics, so having studied any of them will reduce the time needed for the others. Her study times, in hours, are given below.

						Previously read								
	-	P	Q	R	S	PQ	PR	PS	QR	QS	RS	PQR	PQS	PRS	QRS
P	6	-	5	5	4	-	-	-	4	3	3	-	-	-	2
Q	12	11	-	9	10	-	9	9	-	-	7	-	-	5	-
R	5	4	4	-	3	3	-	2	-	2	-	-	1	-	-
S	15	12	13	14	-	10	11	-	12	-	-	7	-	-	-

Which order of study enables her to complete her work in the minimum time?

Exercise 3.3 Jo has just put 50p into a game machine. Each 'go' costs him 20p if he chooses to be the attacker and 10p if he chooses to be the defender. He has time for only three games. The points he can expect to score are given in the table below.

Money in machine	50	40	30	20	10
Points for attacker	15	16	17	19	-
Points for defender	12	10	7	6	3

Using dynamic programming, find the plan which maximises the number of points he can score.

Exercise 3.4 A manufacturer of hot air balloons has the following orders

January	:	1 balloon	March	:	1 balloons
February	:	3 balloons	April	:	2 balloons

Each balloon costs £300 to make and whenever there is production there is a setting up cost of £400, regardless of the number made. Maximum production is three balloons per month. Balloons may be stored at a cost of £100 per month and there is storage capacity for two balloons. If there are no balloons in stock at the start and at the finish of this period, use dynamic programming to find the production schedule to keep the manufacturer's costs to a minimum. Draw a network to represent the problem. Show clearly which node would be omitted if the order for April were reduced to one balloon.

4.4   NOTES

It is possible to solve shortest route problems in networks such as Fig. 3. 2 (from Chapter 3, Investigation 2) by dynamic programming.

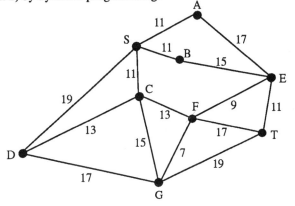

Fig. 4.10

Example

Find the shortest time and the associated route for a journey from S to T.

Solution

Let $d_4(B)$ be the length of the shortest route from B to T using at most four arcs.

Stage 1: $d_1(A) = d_1(B) = d_1(C) = d_1(D) = d_1(S) = \infty$
                     $d_1(E) = 11, d_1(F) = 17, d_1(G) = 19.$

Now consider those nodes linked directly to E, F or G.
                     $d_2(A) = 28, d_2(B) = 26, d_2(C) = 30, d_2(D) = 36,$
                     $d_2(E) = 11, d_2(F) = 17, d_2(G) = 19.$

Examples of nodes
                     $d_2(A) = AE + d_1(E)$
                     $d_2(C) = \min \{CF + d_1(F), CG + d_1(G)\}$
                     $d_2(G) = \min \{d_1(G), GF + d_1(F)\}$

Next look at nodes joined directly to those for which $d_2 > 0$, i.e. A, B, C, D, E, F, G.
                     $d_3(A) = 28, d_3(B) = 26, d_3(C) = 30,$
                     $d_3(D) = 36, d_3(E) = 11, d_3(F) = 17,$
                     $d_3(G) = 19, d_3(S) = 37,$

as $\quad d_3(S) = \min \{SA + d_2(A), SB + d_2(B), SC + d_2(C), SD + d_2(D)\}$

$\qquad\quad = \min \{11 + 28, 11 + 26, 11 + 30, 19 + 36\}$

$\qquad\quad = 37$

The value of $d_3(S)$ depends on that of $d_2(B)$. It may be that a route with more than two arcs from B to T will have a shorter length than $d_2(B)$. For example, if $d_5(B) < d_2(B)$, this will produce $d_6(S) < d_3(S)$ at the next stage. With n nodes, $n-1$ stages of calculations are necessary to be sure of optimisation, so values of d_8 are required for the given network.

As a result, dynamic programming is computationally inefficient when compared with Dijkstra's algorithm for tackling such problems.

5

Flows in networks

5.1 INVESTIGATIONS
Investigation 1
While its bypass is undergoing major reconstruction a town has all the west to east traffic
passing through its centre. In Fig. 5.1 the labels on arcs show the numbers of cars that can
pass along the roads in a five-minute period. Arrows indicate permitted directions in **one-
way** streets and numbers on unarrowed arcs show the capacity in **each** direction.

  What is the maximum west-east flow through the system?

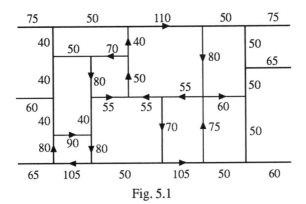

Fig. 5.1

Investigation 2

Taking the labels on the arcs as capacities (i.e. 20 units can flow along arc SA in a given time), find the maximum possible flow through the network from S to T.

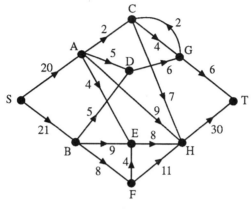

Fig. 5.2

5.2 TERMINOLOGY

A node with arcs carrying flows **from** it only, is called a **source**. Point S in Fig. 5.2 is an example of a source.

A node with arcs carrying flows **to** it only, is called a **sink**. Point T in Fig. 5.2 is an example of a sink.

5.3 CUTSETS

In the network in Fig. 5.3, only 33 units can flow through the arcs SA, SB and SC.

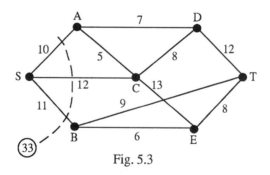

Fig. 5.3

The dotted line forms a **cutset**, that is, it disconnects the network with the source and sink in different parts, and no smaller group chosen from among them will do this. Since only 33 units can flow through here, there can be no more than 33 units of flow possible through the whole system. Is it possible to find a cutset with a lower value?

The cutset (SB, CE, DT) has a value of 36, which tells you nothing new as it places a restriction of 36 on the maximum flow, which you knew could not exceed 33 anyway. However, the cutset (ET, BT, DT) has a value of $8 + 9 + 12 = 29$, so this gives new information about the system's optimal flow.

Can a flow of 29 units be found? One method of achieving this flow is shown in Fig. 5.4.

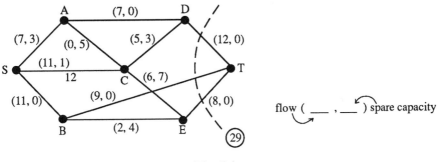

Fig. 5.4

A cutset will always give a maximum possible flow, but you should realise that it may not be easy to find an actual flow as large as the cutset (there may be a smaller cutset which you have not found).

Any feasible flow gives a lower bound for the optimal arrangement, while any cutset provides an upper bound. If flow and a cutset of equal size can be found, then an optimum solution has been achieved.

Notice that the arcs of the minimum cutset (ET, BT, DT) are saturated by the optimal flow, that is, all their capacities are utilised. Other arcs may also be saturated. If a cutset can be found together with a way of saturating its arcs, then a maximum flow will be achieved.

5.4   MAXIMUM FLOW, MINIMUM CUT THEOREM
The **maximum flow** from a source, S, to a sink, T, in a network is equal to the **minimum value** of a cutset disconnecting S from T. If a flow and a cut can be found with the same value, then the flow is a maximum and the cut is a minimum.

5.5   CUTSETS IN DIRECTED NETWORKS
The first cutset considered in section 5.3 had value 33, that is, it was a set of arcs which could carry at most 33 units from left to right.

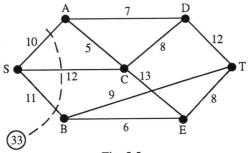

Fig. 5.5

In the directed network in Fig. 5.6, what are the values of the cutsets labelled (a) and (b)?

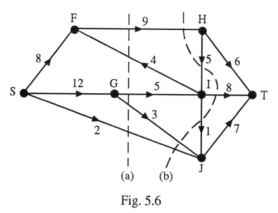

Fig. 5.6

The **maximum** number of units that can flow from left to right through

$$\text{(a) is } 9+0+5+3+2 = 19$$

and through   (b) is $9+0+8+1+3+2 = 23$, as the most that can flow from F to I or I to H is zero.

Now is it possible to find a cutset with a lower value, or should a flow of 19 be sought? A cutset with lower value is (SF, GI, GJ, SJ) which gives 18. A corresponding flow is shown in Fig. 5.7.

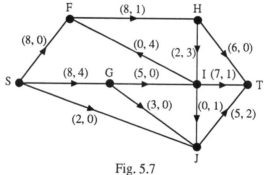

Fig. 5.7

So the maximum possible flow from S to T is 18 units.

5.6   RESTRICTIONS ON NODES

Sometimes there is an upper bound on the flow that can pass through a node, as in Fig. 5.8, where only 14 units can pass through P.

The node concerned may be replaced by a pair of nodes connected by an arc having the value of the restriction on the node, as in Fig. 5.9.

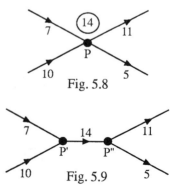

Fig. 5.8

Fig. 5.9

Some care should be taken over the positioning
of arcs in the re-drawn network. The network in
Fig. 5.10 illustrates this point.

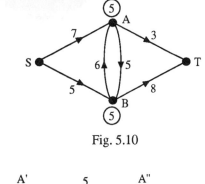

Fig. 5.10

In its revised form, the arcs that run from
A to B and B to A now go from A' to B"
and B" to A'. In the original diagram, flow
could pass from S to A, through the
restriction there, on to B, through the limit
there, and so to T. Therefore the arcs are
re-drawn as shown here.

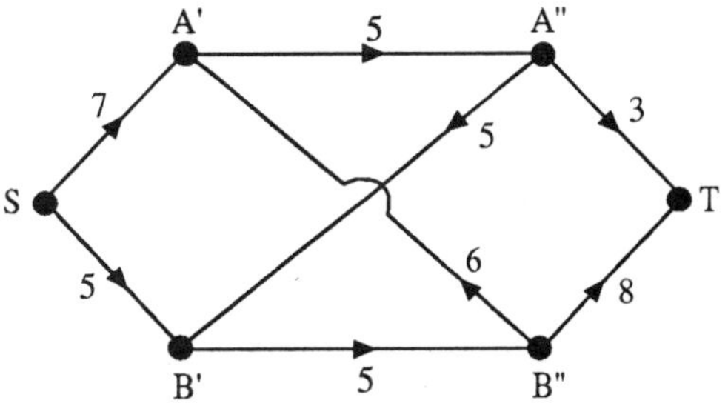

Fig. 5.11

At first sight, the node restriction in Fig. 5.12 (a) may seem a problem, but once the network
is re-drawn with all the 'in flows' on one side and the 'out flows' on the other (Fig. 5.12 (b)),
it can be tackled as before.

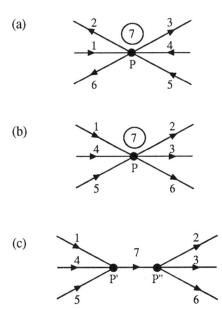

Fig. 5.12

5.7   SEVERAL SOURCES AND/OR SINKS

If a network has several sources, these can be connected to a single **supersource** S, with arcs
of capacity corresponding to the flows from the original sources (Figs. 5.13 and 5.14).

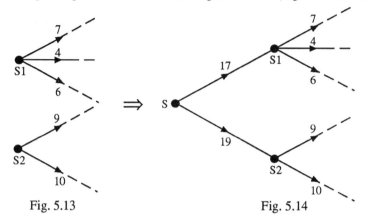

Fig. 5.13   Fig. 5.14

Similarly, several sinks may be connected to a single **supersink.**

With the ideas of sections 5.6 and 5.7, you can always obtain a network with **one source, one
sink** and **all the capacities** on the arcs.

Exercise 7.1   For each of the following networks, find the maximum flow and a minimum
cutset.

(a)

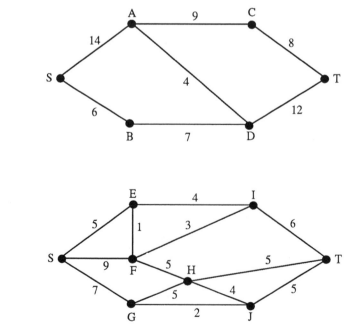

(b)

(c)

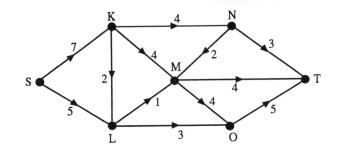

(d)

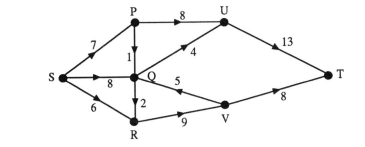

(e)

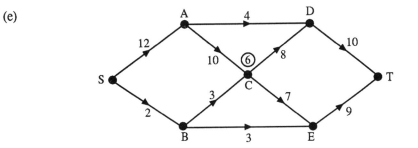

(f)

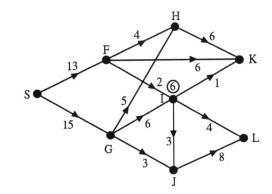

Fig. 5.15

Exercise 7.2   The network shows a road system with capacities in cars per hour.

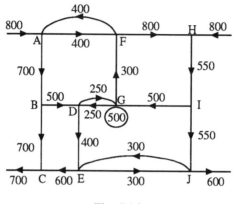

Fig. 5.16

Redraw the diagram with a single source, a single sink, and all the capacities on arcs. What is the maximum flow per hour through the system?

5.8   FLOW AUGMENTATION SYSTEM

What should you do when the smallest cutset you can find has value 20, for example, and the best flow is only 18? One plan is to check more cutsets, but while the number to look at is certainly finite, it may be very large. Another approach, which is the one which will be considered here, is to try to improve the flow already found.

A flow augmentation algorithm is designed to build on a given flow through a network and can be applied repeatedly until the optimum situation is reached.

An algorithm which labels the nodes in succession from source to sink will be used.

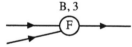

The label above node F shows that it has been reached from B and that it is possible to send an additional flow of 3 units from there.

Each arc will be labelled as in section 5.3, so Fig. 5.17 shows a flow of 4 from C to E and the ability of the arc to carry a further 1 unit if required.

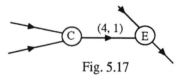

Fig. 5.17

Example

The network in Fig. 5.18 can carry a flow of 15 by sending 6 through SAT, 3 through SBDT and 6 through SCT. Show how this flow may be augmented to give a maximum flow of 20.

Solution

Send 8 more units along SC and label C with 5, 8. CT is saturated but 5 of the units can be sent along CD, so D is labelled C, 5. Only 2 of these units can now be routed along DT to T, so T is labelled D, 2 and the flow has been augmented by 2 units (Fig. 5.19).

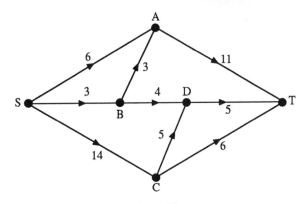

Fig. 5.18

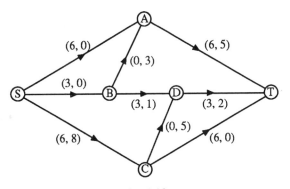

Fig. 5.19

The complete labelling is shown in Fig. 5.20, and now the network can be drawn with the new flow shown (Fig. 5.21).

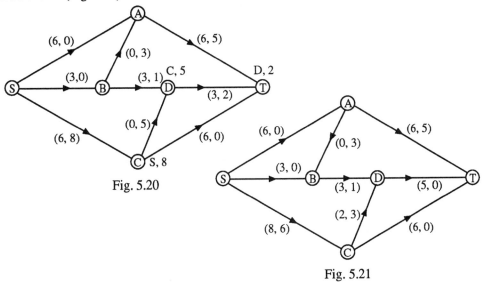

Fig. 5.20

Fig. 5.21

Now try to increase the flow further. Send six units from S to C, which is labelled S, 6. Three of these units can be sent to D, which receives the label C, 3. Now none of these units can leave D, as the only arc which can carry an out-flow is saturated. However, this need not be the end of your attempt if some of the flow arriving at D from elsewhere can be re-routed. You need to send the three units that have arrived at D (from C) along DT to the sink, so reduce the flow along BD by three units and label B with –D, 3. The three units now at B that can no longer go along BD can be sent to A which is, therefore, labelled B, 3. There is spare capacity in AT so the three units can pass on to T, which is labelled A, 3. The total flow is now 20 units. The network and labelling are shown in Fig. 5.22.

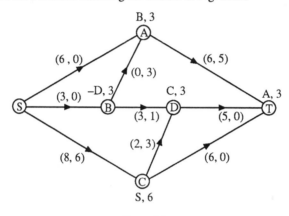

Fig. 5.22

The diagram is now redrawn (Fig. 5.23) to show the increased flow of 20 units, but is this optimal?

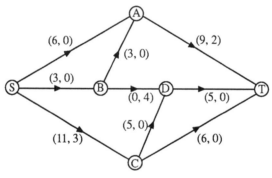

Fig. 5.23

Three units may be sent to C, which is labelled S, 3. However, there is no spare capacity in the arcs leaving C nor is it possible to re-route any of the flow coming to it. The process therefore comes to an end as there is no other way to send more units from S. The flow of 20 units is the maximum possible. (Note that SA, BA, DT, CT form a saturated cutset.)

The following algorithm uses the two ideas of sending more units along unsaturated arcs in steps 2a and 3a, and the re-routing of existing flow in 2b and 3b.

Step 0:   Give each arc a feasible flow, making sure that the flow is conserved
           at each node.

Step 1:   Label the source $(-,\infty)$.

Step 2:   Scan the arcs until one is found for which
           (a)   the tail node is labelled while the head node is not, and the arc
                 has spare capacity,

           or

           (b)   the head node is labelled while the tail node is not and there is
                 already a positive flow along the arc.

           If no such arcs exist, go to step 5.

Step 3:   (a)   Label the head node with (tail, MIN of label at the tail and the
                 extra capacity).

           (b)   Label the tail node with (–head, MIN of label at the head and
                 the flow in the arc).

           If the sink is labelled, go to step 4. Otherwise, go back to step 2.

Step 4:   A complete chain has been found from the sink to the source and the flow
           can be augmented by the label at the sink.

           Go to step 2.

Step 5:   The optimal flow has been found. **Stop!**

Example

Step 0

Assign a feasible flow of 6.
3 in SBET and 3 in SACDT

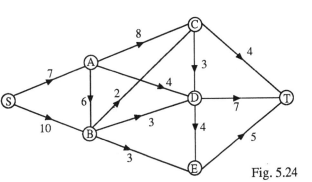

Fig. 5.24

Step 1
Label S with –, ∞.

Steps 2, 3
Label A-S, 4; D-A, 4; T-D, 4.

Step 4
Augment flow by 4 along SADT
to give new total flow of 10.

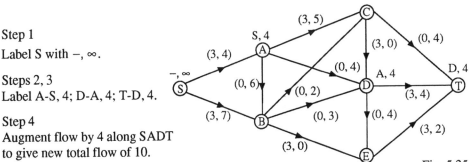

Fig. 5.25

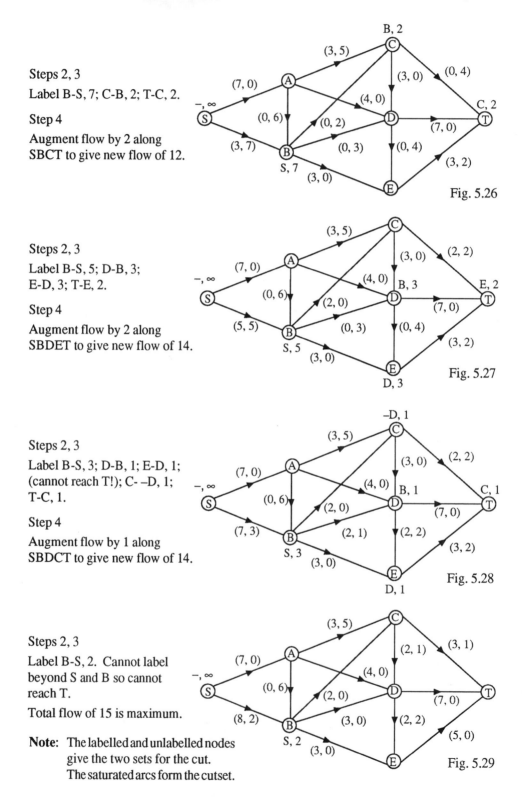

Steps 2, 3

Label B-S, 7; C-B, 2; T-C, 2.

Step 4

Augment flow by 2 along
SBCT to give new flow of 12.

Fig. 5.26

Steps 2, 3

Label B-S, 5; D-B, 3;
E-D, 3; T-E, 2.

Step 4

Augment flow by 2 along
SBDET to give new flow of 14.

Fig. 5.27

Steps 2, 3

Label B-S, 3; D-B, 1; E-D, 1;
(cannot reach T!); C- –D, 1;
T-C, 1.

Step 4

Augment flow by 1 along
SBDCT to give new flow of 14.

Fig. 5.28

Steps 2, 3

Label B-S, 2. Cannot label
beyond S and B so cannot
reach T.

Total flow of 15 is maximum.

Note: The labelled and unlabelled nodes
give the two sets for the cut.
The saturated arcs form the cutset.

Fig. 5.29

Exercise 8.1 Use the flow augmentation (labelling) algorithm to find the maximum possible flow through the following networks.

(a) In Fig. 5.30, start with a feasible flow SBACT = 2, SBDT = 3.

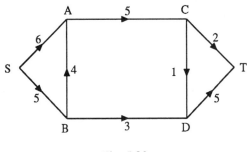

Fig. 5.30

(b) In Fig. 5.31, start with a feasible flow SAT = 2, SBT = 1, SCT = 2.

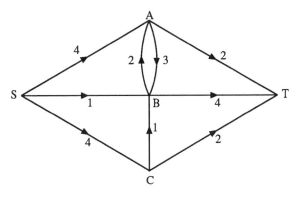

Fig. 5.31

(c) In Fig. 5.32, start with a feasible flow SACET = 10, SBDFT = 15, SBDFET = 12.

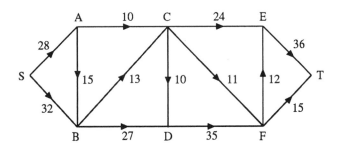

Fig. 5.32

Exercise 8.2   Adjust the following networks in suitable ways and find the maximum possible flows from sources to sinks.

(a)

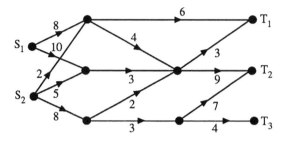

Fig. 5.33

(b)

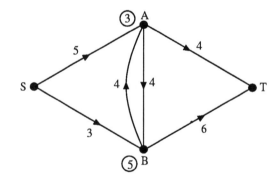

Fig. 5.34

(c)

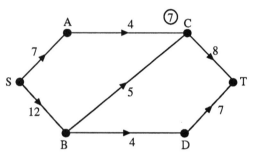

Fig. 5.35

5.9   NETWORKS WITH ARCS HAVING LOWER CAPACITIES

Beware of confusion over notation similar to that used before.

5.9.1 Max-flow, min-cut theorem

The arcs in the network in Fig. 5.36 each have two numbers on them – the lower and the upper capacities of those arcs, respectively.

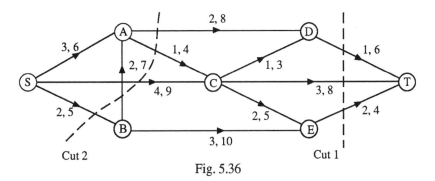

Fig. 5.36

The lower capacity is the minimum flow that must be at least equalled. So for the arc AD, the flow along it must satisfy $2 \le$ flow ≤ 8.

$$\underset{A}{} \xrightarrow{\quad 2,\,8 \quad} \underset{D}{}$$

In the network in Fig. 5.36,   capacity of cut $1 = 6 + 4 + 8 = 18$;

capacity of cut $2 = 8 + 4 + 9 - 2 + 5 = 24$.

That is, for flows from the left, use the maximum values and for those from the right, subtract the minimum values.

The arc $\xrightarrow{-7,\,-2}$ could replace the $\xleftarrow{2,\,7}$ in the cut and the maximum values taken on each arc, so producing the -2 again.

All this is consistent with section 5.6.

5.9.2   Flow augmentation algorithm

The labelling algorithm can be applied to networks with lower capacities. Finding an initial feasible flow is more difficult as all the lower restrictions must be satisfied. (Before it was possible to start with a zero flow in each arc which would suit computers very well.) In addition when step 2b is employed to reduce a flow in an arc, care must be taken not to go below the lower capacity.

Example

A feasible flow is
SACT 5, SADT 1, SABDT 2,
SBDT 1 so total flow is 9.

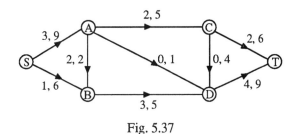

Fig. 5.37

Steps 2, 3

Label B-S, 5; D-B, 2; T-D, 2.

Step 4

Augment flow by 2 in

SBDT to give new flow of 11.

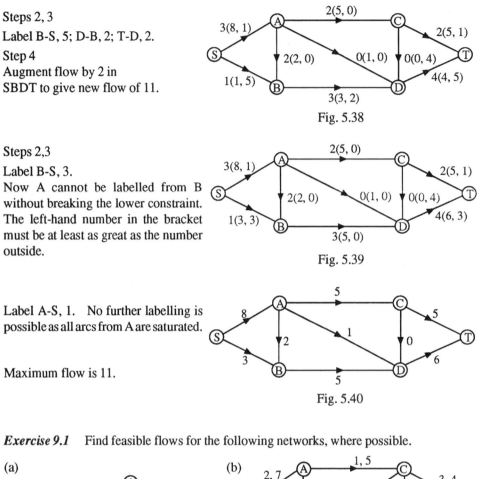

Fig. 5.38

Steps 2,3

Label B-S, 3.

Now A cannot be labelled from B
without breaking the lower constraint.
The left-hand number in the bracket
must be at least as great as the number
outside.

Fig. 5.39

Label A-S, 1.   No further labelling is
possible as all arcs from A are saturated.

Maximum flow is 11.

Fig. 5.40

Exercise 9.1   Find feasible flows for the following networks, where possible.

(a)

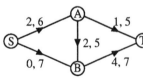

Fig. 5.41

(b)

Fig. 5.42

(c)

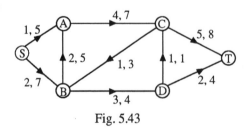

Fig. 5.43

Exercise 9.2   Find the minimum cut for each of the networks in Exercise 9.1.

Exercise 9.3   Find a feasible flow for the network in Fig. 5.44, and use the flow augmenting algorithm to find the maximum flow.

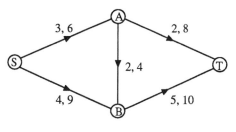

Fig. 5.44

Exercise 9.4   Show that the max-flow, min-cut theorem holds for the network in Fig. 5. 45.

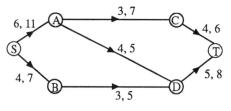

Fig. 5.45

Exercise 9.5   By considering the arcs at A in Fig. 5.46, give the minimum value for *b* possible for a feasible flow. What do the arcs at D tell you about *a*?

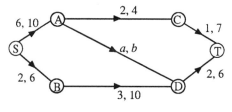

Fig. 5.46

5.10 NOTES

In all the network flow problems it is not easy to find a cut which gives the optimum flow, and you are left wondering if a better cutset can be found. The following example describes an approach which will help to determine a good solution.

In Fig. 5.47, treat the numbers on the arcs as distances and find the shortest distance from S to T. (39)

Use the numbers on the arcs as maximum capacities and find the optimal flow from S to T. (43)

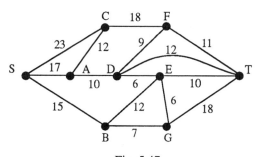

Fig. 5.47

Construct the **dual** network as shown in Fig. 5.48.

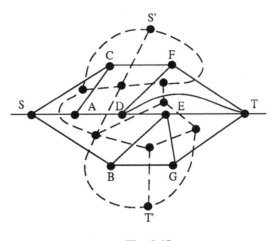

Fig. 5.48

A node has been placed in each of the original regions and each arc has been cut exactly once by an arc of the dual.

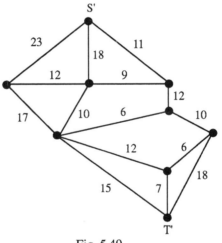

Fig. 5.49

Each arc in the dual has the same label as the one in the original it crosses. Treat the numbers on the arcs as maximum capacities and find the optimal flow from S' to T'. (39)

Use the numbers on the arcs as distances and find the shortest distance from S' to T'. (43)

So a maximum flow question can be connected to one about shortest distance, which may be tackled using methods such as Dijkstra's algorithm.

6

Critical path analysis

6.1   ACTIVITY ON ARC :   INVESTIGATIONS
Investigation 1

The network in Fig. 6.1 shows the activities involved in preparing a meal and the times, in minutes, for each activity.

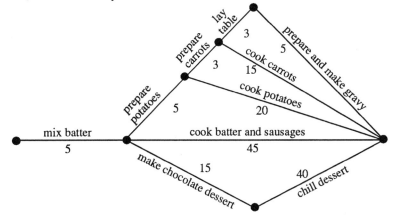

Fig. 6.1

(a)   What is the shortest possible time required to prepare the meal?
(b)   How many people are involved in the preparation?
(c)   How may the total time be reduced?
(d)   Can people start to eat after 50 minutes?

Investigation 2

Draw up a network for planning a sponsored walk, to include the following activities.

A: Find a date

B: Plan a route with the police

C: Print forms

D: Get walkers

E: Allow walkers time to gather sponsors

F: Organise refreshments

G: Contact St. John Ambulance for first aid provision

H: Organise transport for those unable to finish.

Add any other activities you think necessary and estimate the times needed for each one.

6.2   ACTIVITY ON ARC :   DEFINITIONS AND CONVENTIONS

An **activity** is an element of the **work needed** for the completion of the project. (e.g. waiting for cement to harden). In the first four sections of this chapter, each activity is represented by an arc with an arrow, hence the title 'activity on arc'. There is another convention which you will meet later in the chapter.

Event on node

An **event** is the start or finish of an activity or group of activities. Events are represented by numbers, usually within circles.

Conventions

(a)   Time flows from left to right.

(b)   Head nodes always have a higher number than those of associated tail nodes.

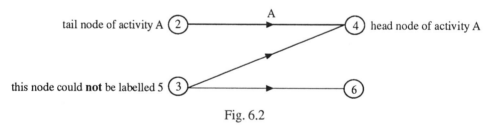

Fig. 6.2

Merge

Events into which more than activity enter are known as **merge nodes** (Fig. 6.3).

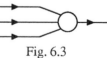

Fig. 6.3

Burst

Events which have more than one activity leaving are known as **burst nodes** (Fig. 6.4).

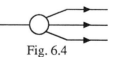

Fig. 6.4

There is no reason why an event should not come
into both categories of merge and burst (Fig. 6.5).

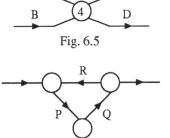

Fig. 6.5

Loops
Loops cannot exist if the convention about time
flowing from left to right is adhered to.

Rather than use the network in Fig. 6.7, it is preferable
to draw the network in Fig. 6.8.

Fig. 6.6

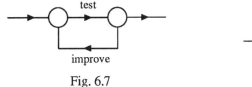

Fig. 6.7

test and improve

Fig. 6.8

Dummy activities
Some situations require the use of activities that consume neither time nor resources; these
are termed **dummy activities**. These are employed to:

(a) ensure that the identities of distinct activities really are different;

(b) avoid errors in logic.

Dummy activities are usually represented by dotted
lines, but activities labelled D or 0 are also seen in
some texts.

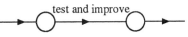

(a) Activities A and B in Fig. 6.9, have the same
head and tail nodes.

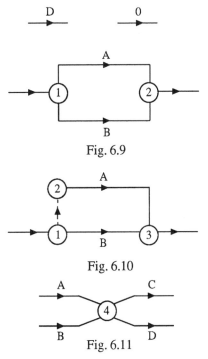

Fig. 6.9

The introduction of a dummy activity in
Fig. 6.10 avoids any loss of identity. Each
activity can be uniquely described in terms
of its head and tail nodes.

$$A = (2, 3) \qquad B = (1, 3)$$

Fig. 6.10

(b) If A precedes C, and A and B precede D,
then the diagram in Fig. 6.11 is faulty.

This representation is incorrect as it shows
C preceded by both A and B.

Fig. 6.11

The difficulty is overcome by introducing a
dummy activity as shown in Fig. 6.12.

Dummy activities **must** have direction.

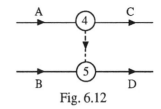

Fig. 6.12

Always examine any 'crossroads' that occurs when a network is drawn, to check that
dependence is correctly represented.

Overlapping activities

It is not always true that an activity need stop for a succeeding one to begin. A sports writer
may well start to write a report before the event is completed, so instead of

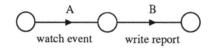

watch event   write report

Fig. 6.13

if the writer starts after half an hour of a ninety-minute event, the diagram becomes that shown
in Fig. 6.14.

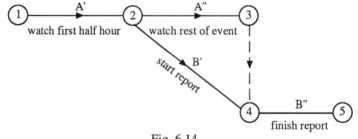

Fig. 6.14

Information at nodes (event times)

In the following section, the information indicated at each node is given.

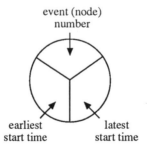

Fig. 6.15

6.3   ACTIVITY ON ARC : PROCEDURE

You can establish earliest start times for each activity by working forwards through the network and then finding the latest start time by a backward pass. The **critical path** is determined by passing through the nodes that have equal values for the earliest start times and latest start times.

Example

A project consist of activities A to N. Their durations and precedences are shown in the following table.

Activity	Duration (hours)	Precedents
A	8	–
B	15	A
C	5	B
D	7	A
E	13	A
F	2	D
G	3	C, E, F
H	2	D
I	5	H
J	2	G, H
K	4	J
L	2	K
M	3	I, K
N	4	L, M

(a) Draw the network diagram.

(b) Determine the earliest start time (EST) and the latest start time (LST) at each node and so find the critical path.

(c) If the start of C is delayed by one hour for lack of materials, how will this affect the total time required to complete the project?

Solution

(a)

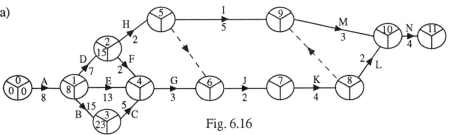

Fig. 6.16

[The diagram could be redrawn so that the dummy 8–9 is directed from left to right, but since zero time flows along this arc there is no problem with convention.]

The numbering of nodes is rarely unique. For example, in Fig. 6.16 the labels two and three could easily change nodes.

(b) **Forward pass**

At each node the earliest start time is the **maximum** of any possible

earliest start time **plus** duration of activity

for each node directly **preceding** it.

For example, for node 4,

EST = maximum of EST of 2 + duration of F,

EST of 1 + duration of E,

EST of 3 + duration of C

= maximum of (15 + 2, 8 + 13, 23 + 5)

= maximum of (17, 21, 28)

= 28

That is, after 28 days activity, G can start as soon as C, E and F are completed.

For node 9,

EST = maximum of (EST of 5 + duration I, EST of 8 + 0)

= maximum of (17 + 5, 37 + 0)

= 37

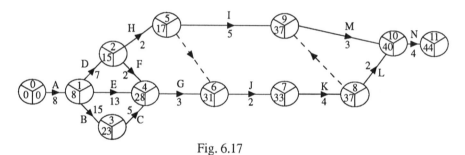

Fig. 6.17

Backward pass

At each node the latest start time is the **minimum** of

latest start time **minus** duration of activity

for each node immediately **following** it.

For example, for node 8,

LST = minimum of (LST of 10 – duration of L, LST of 9 – duration 0)

= minimum of (40 – 2, 37 – 0)

= minimum of (38, 37)

= 37

For node 2,

LST $=$ minimum of (LST of 5 – 2, LST of 4 – 2)

$=$ minimum of (31 – 2, 28 – 2)

$=$ minimum of (29, 26)

$=$ 26

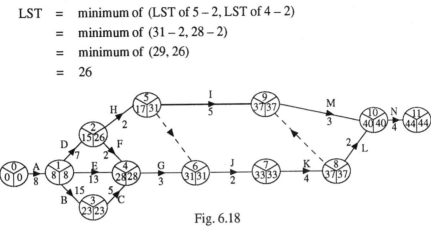

Fig. 6.18

The critical path is made up of those events which determine the **earliest** possible completion time of the project. The critical path can be shown with double lines when it has been found.

(c) C is on the critical path and therefore is called a **critical activity**. Any delay affecting it will increase the total time for the project so the completion time is now 45 hours.

Remember that an activity is **critical** if EST = LST at both the head and tail nodes. Note also that the **duration** of an activity = EST at head – EST at tail.

Exercise 3.1 Draw the following network

Activity	A	B	C	D	E	F	G	H	I
Precedents	–	A	B	A	D	C, E	D	–	F, H

Exercise 3.2 Draw the following network. (Hint: use an identity dummy.)

Activity	A	B	C	D
Precedents	–	–	A, B	C
Duration (days)	5	10	8	4

Show the earliest and latest start times at each node and mark on the critical path. If A is delayed by six days due to bad weather, how will the total time for the project be affected?

Exercise 3.3 A project consists of activities A to G as detailed in the table. (Hint: use a logic dummy.)

Activity	A	B	C	D	E	F	G
Precedents	–	–	B	A	D	C, D	E, F
Duration (days)	6	3	5	8	1	5	2

Draw the network showing

(a) EST and LST at each node;   (b) the critical path;

(c) the total time needed to complete the project.

List the non-critical activities and give the time each one may be delayed without changing the overall project time.

Exercise 3.4   A project comprises activities A to K with details as shown.

Activity	A	B	C	D	E	F	G	H	I	J	K
Precedents	–	–	A	A	D	E	C, D	G, F	E	G, I	J, H
Duration (hours)	4	11	1	4	1	3	7	5	2	1	4

(a) Draw the network.

(b) Determine the earliest and latest starting times at each node.

(c) Find the duration of the project and list the critical activities.

(d) Show how you would alter the network if D can start after A has been in progress for two hours.

Exercise 3.5   A project is made up of activities A to K as given below.

Activity	A	B	C	D	E	F	G	H	I	J	K
Precedents	–	–	see below	A	B	A, C	F	D	D	H, I	D, E, G
Duration (hours)	3	5	4	4	3	1	2	2	1	3	6

Draw the network and find the duration of the project and the critical activities if C can start two hours after B has begun and cannot finish before B.

Exercise 3.6   Draw the network to describe the following project and find the critical paths.

Activity	A	B	C	D	E	F	G	H	I	J
Precedents	–	–	–	A	A B C	C	D E	E	F G H	F G H
Duration (days)	5	4	5	7	4	7	2	5	6	3

(a) What is the duration of the project?

(b) Which activities are not critical?

6.4 ACTIVITY ON NODE : INVESTIGATIONS
Investigation 1

The network in Fig. 6.19 shows the activities involved in preparing a meal and the times in minutes for each.

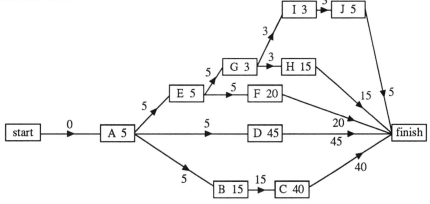

Fig. 6.19

A :	Mix batter	F :	Cook potatoes
B :	Make chocolate dessert	G :	Prepare carrots
C :	Chill dessert	H :	Cook carrots
D :	Cook batter and sausages	I :	Lay tables
E :	Prepare potatoes	J :	Prepare and make gravy

(a) What is the shortest time required to prepare the meal?

(b) How many people are involved in the preparation?

(c) How may the total time be reduced?

(d) Can people start to eat after 50 minutes?

Investigation 2

Draw up a network for planning a sponsored walk. Include the following activities.

    A : Find a date
    B : Plan a route with the police
    C : Print forms
    D : Get walkers
    E : Allow walkers time to gather sponsors
    F : Organise refreshments
    G : Contact St. John Ambulance for first aid provision
    H : Organise transport for those unable to finish

Add any other activities you think necessary and estimate the times needed for each of the activities involved.

6.5   DEFINITIONS AND CONVENTIONS

Activity on node: this is an alternative convention to that already used. Activities are represented by **nodes** and are drawn as **rectangles**.

Dependent activities

Fig. 6.20 shows an activity A, which takes 7 minutes to complete and a second activity B, which must come after it. The number on the arc is the time necessary between the starts of the two events, so, in this case, B may begin as soon as A is completed.

Fig. 6.20

Overlapping activities

It may be necessary for an activity to be completed before a dependent one can be started. In the activity represented in Fig. 6.21, D may begin three minutes after C has started. Notice that the completion times for C and D are six and eight minutes after the start of C respectively.

Fig. 6.21

Now F's finishing time comes before E's and whether this is possible or not depends on what the activities actually are (Fig. 6.22).

Fig. 6.22

If F must have its completion after E's, then the activities can be split so that E'= start E, and E" = finish E. Now F can start but not be completed before E is finished (Fig. 6.23).

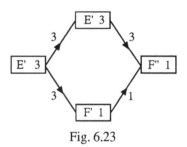

Conventions

(a)   Time flows from left to right so all dependency arrows point to the right.

(b)   When nodes are labelled with numbers, the number at the head of an arrow is always greater than the number at the tail.

Fig. 6.23

Merge

An activity such as I in Fig. 6.24, which has at least two immediately preceding activities is called a **merge**.

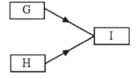

Fig. 6.24

Burst

An activity, such as J in Fig. 6.25, which has at least two immediately dependent activities is called a **burst**.

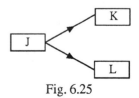

Fig. 6.25

Loops

Loops cannot exist in a network as they contravene the convention that dependency arrows point from left to right (Fig. 6.26).

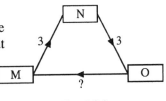

Fig. 6.26

Information at nodes

The information indicated at each node will be given (Fig. 6.27).

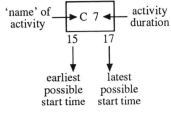

Fig. 6.27

6.6 ACTIVITY ON NODE : PROCEDURE

Establish the earliest start times by a forward pass through the network and then find the latest start times by a backward pass. Show the critical path clearly.

Example

A project consists of activities A to N; their durations and precedences are shown below.

Activity	Duration (hours)	Precedents
A	8	–
B	15	A
C	5	B
D	7	A
E	13	A
F	2	D
G	3	C, E, F
H	2	D
I	5	H
J	2	G, H
K	4	J
L	2	K
M	3	I, K
N	4	L, M

Solution

(a) Draw the network diagram.

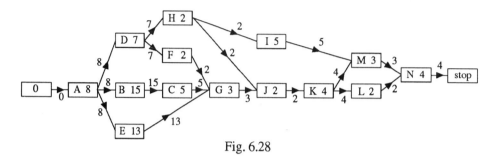

Fig. 6.28

(b) Determine the earliest start time and latest start time at each node and so find the critical path.

Forward pass

At each node the earliest start time is the maximum of

earliest start time + arc time

for each node directly preceding it.

For node G,

$$\begin{aligned} EST &= \text{max} \quad (EST\ C + 5, EST\ F + 2, EST\ E + 13) \\ &= \text{max} \quad (23 + 5, 15 + 2, 8 + 13) \\ &= \text{max} \quad (28, 17, 21) \\ &= 28 \end{aligned}$$

That is, after twenty eight hours activity G can start as C, E and F are complete (Fig. 6.29).

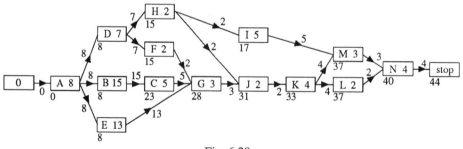

Fig. 6.29

Backward pass

At each node the latest start time is the minimum of

latest start time – arc time

for each node immediately following it.

For node K,

$$\text{LST} = \min \quad (\text{LST } M - 4, \text{LST } L - 4)$$
$$= \min \quad (37 - 4, 38 - 4)$$
$$= \min \quad (33, 34)$$
$$= 33$$

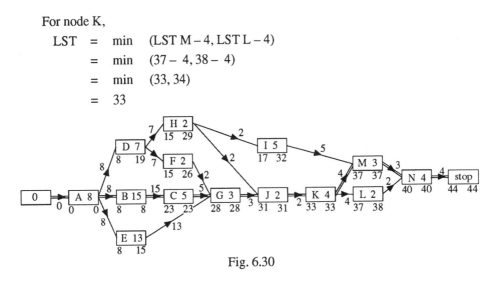

Fig. 6.30

The critical path is made up of those activities which determine the earliest possible completion time of the project and is shown as double lines (Fig. 6.30).

(c) If the start of C is delayed by one hour for lack of materials, how will this affect the total time required to complete the project?

C is on the critical path and is, therefore, call a **critical activity**. Any delay affecting it will increase the total time for the project as the completion time is now 44 hours.

An activity is critical if EST = LST and the critical path connects critical activities along arcs for which

head node EST – tail node EST = arc time.

Exercises 6.1 to 6.6 Re-work Exercises 3.1 to 3.6, ignoring references to logic and identity dummies.

6.7   ACTIVITY ON NODE : TOTAL FLOAT

A non-critical activity possesses some flexibility in its starting time.

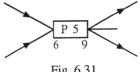

Fig. 6.31

The total float = latest start time – earliest start time
                  = 9 – 6 = 3 units,

and this is the amount P can move its starting time from its earliest value of six without altering the overall project duration.

The total float for any critical activity is zero.

6.8   ACTIVITY ON ARC : TOTAL FLOAT

A non-critical activity possesses some flexibility in its starting time.

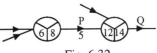

Fig. 6.32

For the activity P in Fig. 6.32, the time available is at most $14 - 6 = 8$ as it must start no earlier than 6 units after the beginning of the project, and be completed by 14 units for Q to get under way by its latest start time. However, it only actually needs five units to be carried out, so its

$$\text{total float} = 8 - 5 = 3 \text{ units.}$$

This is the amount P can move its starting time from its earliest value of six, without altering the overall project duration.

For any activity,

total float = LST of head node – EST of tail node – duration of activity.

The **total float** for any critical activity is **zero**.

6.9   GANTT CHARTS
The project network studied in section 6.3 can be transferred on to a **Gantt chart** showing activities as horizontal lines or bars.

Non-critical activities	Total floats
D	$26 - 8 - 7 = 11$
E	$28 - 8 - 13 = 7$
F	$28 - 15 - 2 = 11$
H	$31 - 15 - 2 = 14$
I	$37 - 17 - 5 = 15$
L	$40 - 37 - 2 = 1$

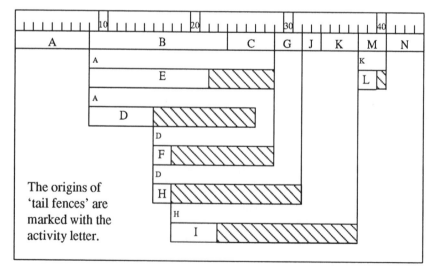

The origins of 'tail fences' are marked with the activity letter.

Fig. 6.33

Looking again at the calculation of H's total float,

$$\underset{\uparrow}{\text{total float}} \ = \ \overset{\overset{\text{end of bar}}{\downarrow}}{31} \ - \ \overset{\overset{\text{unshaded length}}{\downarrow}}{15} \ - \ \underset{\underset{\text{start of bar}}{\uparrow}}{\overset{\text{}}{}} 2 \ = \ \underset{\underset{\text{shaded length}}{\uparrow}}{14}$$

The bar for H can be thought of as shaded all the way from 15 to 31 with actual activity H (two units long) placed on it. This two unit length may be slid along the bar into any position without affecting the overall project time.

There are three vertical 'fences' associated with activity H. The one at 15 indicates that it depends on D and, as the latter slides along its bar F, and H are moved to the right by the same amount, so reducing their total floats. The fence connecting H to I shows that H must be completed before I can commence, and the fence at 31 gives the time by which H must be completed so J can begin. Check that the dependencies given in the original table, or the network, are all shown in the chart.

If a bar of activity ran from 15 to 31 and it had been drawn between F and H, it would have touched the two fences there. However, it may have been on a different part of the network not requiring D's completion, so the fence would be inappropriate. To avoid this sort of problem, some bars can be placed above those of the critical activities.
There is much to commend the drawing of a rough draft.

Each activity can be shown on a separate horizontal line but the form given shows the critical activities clearly and provides a compact, easily viewed picture of the project's details.

The chain of activities K and L has a float of three, but any of this that is utilised by K reduces the amount available to L.

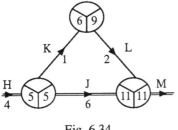

Fig. 6.34

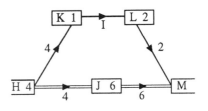

Fig. 6.35

This may be shown in a single bar (Fig. 6.36).

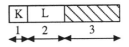

Fig. 6.36

Exercise 9.1   Draw a Gantt chart for the project in Exercise 3.2.

Exercise 9.2   Draw a Gantt chart for the project in Exercise 3.3.

Exercise 9.3   Draw a Gantt chart for the project in Exercise 3.4 (original network).

Exercise 9.4   Draw a Gantt chart for the project in Exercise 3.5.

6.10 RESOURCE LEVELLING

Non-critical activities do not have fixed starting times, so how is the decision when to commence them made? Ideally the project planner would not want to run two activities, each with high manpower requirements, simultaneously. The machinery available may provide a further constraint.

For the Gantt chart in section 6.5, add worker requirements as follows:

A 10	B 7	C 6	D 4	E 2
F 5	G 6	H 7	I 5	J 4
K 6	L 5	M 7	N 2	

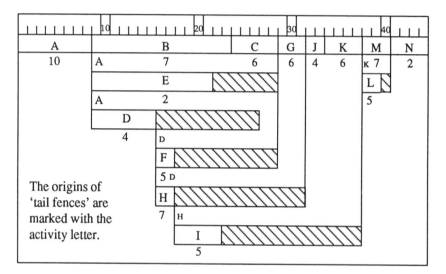

Fig. 6.37

If H moves to hours 30 and 31, the requirements there will be 13, and the highest will be 14 in the period from the 15th to the 21st hour. I has the greatest float of those in the period and could move to hours 33 to 37. This leaves a requirement for 14 for two hours when B, E and F could all be in operation. F has the largest float and can move to hours 24 and 25. This gives the new Gantt chart in Fig. 6.38 and the block graph in Fig. 6.39.

A block graph (Fig. 6.39) to show the number of workers required on each hour, if every activity commences at its earliest start time, shows a situation that would be unacceptable to any planner. Employing 21 people to be able to accommodate all the activities scheduled for hours 16 and 17, and having at least one third of the workforce idle for the rest of the time, is unlikely to be economic.

When should the non-critical activities start if the requirement for personnel is to be made as uniform as possible? There is no algorithm available to calculate the optimal starting times of those activities on non-critical paths. Only heuristic (trial and error) methods exist to help solve the problem. When activities are in competition for resources, priority should be given to satisfying those with least float. Those with most float are more likely to be movable away from the problem times.

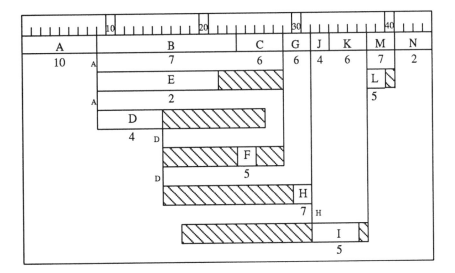

Fig. 6.38

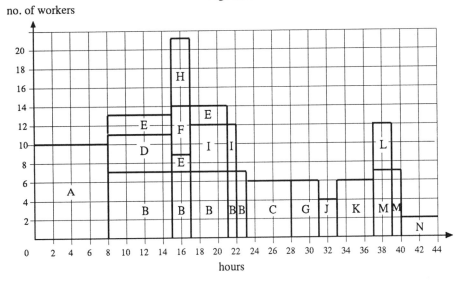

Fig. 6.39

Where B, E, F and H are all needing workers (i.e. at 16 and 17 hours after the start), B's requirements should be met first as it has no float and H is the immediate candidate for a later start as it has the greatest flexibility of timing.

The project can now be undertaken and completed in the time allowed with a workforce of 13 (Fig. 6.40). Whether further improvements can be made may depend on questioning the original logic underlying the problem. For example:

(a)   Could H be done in four hours by four people instead of in two hours by seven people?

(b)   Can an activity be started, left (i.e. some of its float in the middle) and completed later?

(c)   Would it be economic to delay the completion of the project to allow D to follow E?

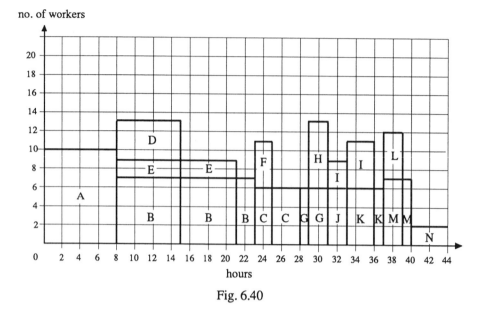

Fig. 6.40

Exercise 10.1   For each of the Gantt charts in the Exercises in section 9, level the resource requirements as much as possible using the following activity needs.

(a)   A : 4   B : 7   C : 8   D : 5

(b)   A : 7   B : 4   C : 3   D : 2   E : 3   F : 4   G : 4

(c)   A :12   B : 2   C : 7   D : 5   E : 2   F : 5   G : 5
      H : 6   I : 4   J : 4   K : 9

(d)   A : 4   B : 7   C : 2   D : 3   E : 2   F : 6   G : 7
      H : 8   I : 7   J : 8   K : 4

6.11   NOTES

The use of start-to-start times in the activity on node section is the **method of potentials** and has some advantages over using durations to label the arcs.

Which is 'better', *activity on arc* or *activity on node?*

As dummies are not needed to sort out identity or logic in activity on node diagrams, perhaps they are easier to put together. However, while the result is likely to be correct, it may well be more complicated to trace through than its activity on arc counterpart. For example, if E, F, G, H, each depend on A, B, C, D, the activity on arc diagram would be as shown in the diagram in Fig. 6.41, while activity on node gives the diagram in Fig 6.42.

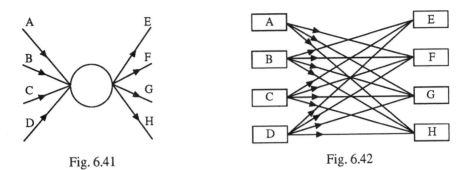

Fig. 6.41   Fig. 6.42

This is where activity on node may introduce a dummy activity to simplify the diagram if overlaps allow.

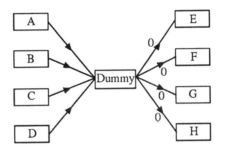

Fig. 6.43

The Gantt or cascade chart is a simple and effective way to display a project but its drawback is in showing clearly the interdependencies that exist between the various activities.

7

Linear programming (graphical)

7.1 INVESTIGATION

The common room is to be extended and some new chairs will be bought as part of the furnishing improvements. The easy chairs are obtained from a firm who can offer two types; chairs with arms which cost £25 each and chairs without arms which cost £20 each. The sum of £2200 has been put aside to buy chairs, and at least one hundred must be purchased.

(a) What is the greatest number of chairs that can be bought?

(b) What is the maximum number of chairs with arms that can be purchased?

(c) What is the cheapest arrangement?

(d) If at least one chair with arms is bought for each four without, how will this change the answers to the questions above?

7.2 GRAPHICAL REPRESENTATION
Example

A firm has to move 1200 parcels in its lorries and vans. Each lorry can carry 200 parcels and each van can take 50 parcels. There are twelve drivers available and seven lorries and fifteen vans. How many different arrangements of lorries and vans can do the job?

If it costs £75 to use a lorry and £25 to use a van, what is the cheapest way to do the job?

Solution

What can be varied?

The number of lorries (call it x) and the number of vans (call it y) can be varied; these are called the **variables**.

What is the objective?

To do the delivery as cheaply as possible. The cost is $75x + 25y = c$. The aim is to find the minimum value of c and the corresponding values of x and y. The function $c = 75x + 25y$ is called the **objective function**.

What constraints restrict the choices of values for the variables?

There are only 12 drivers available, so	$x + y \le 12$	(1)
You can use at most seven lorries, so	$x \le 7$	(2)
You can use at most 15 vans, so	$y \le 15$	(3)

The lorries and vans must have a total capacity of at least 1200 parcels, so $200x + 50y \ge 1200$ which can be simplified by dividing by 50 to give

$$4x + y \ge 24 \tag{4}$$

These four inequalities are called the **constraints**. $x \ge 0$ and $y \ge 0$ are constraints, but are trivial.

Each of the four constraints can be shown as a region on a graph. $x \le 7$ is the area to the left of the vertical line $x = 7$ and the line itself. This is shown by drawing $x = 7$ as a solid line and shading the **unwanted** region on the right (Fig. 7.1).

The line $x + y = 12$ passes through the points (0, 12) (12, 0) (6, 6) so it too can be drawn. Choose a point on one side of the line, e.g. (0, 0). This satisfies $x + y \le 12$ as $x + y = 0 + 0$ here so it is in the required region. Shade the other side of the line (Fig. 7.2).

The line $4x + y = 24$ passes through the points (6, 0) (0, 24) (3, 12) so it can be drawn as shown (Fig. 7.3). Choose a point on one side of the line, e.g. (0, 0). This does not satisfy $4x + y \ge 24$ so the side containing (0, 0) is shaded.

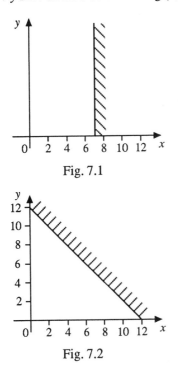

Fig. 7.1

Fig. 7.2

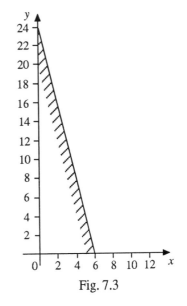

Fig. 7.3

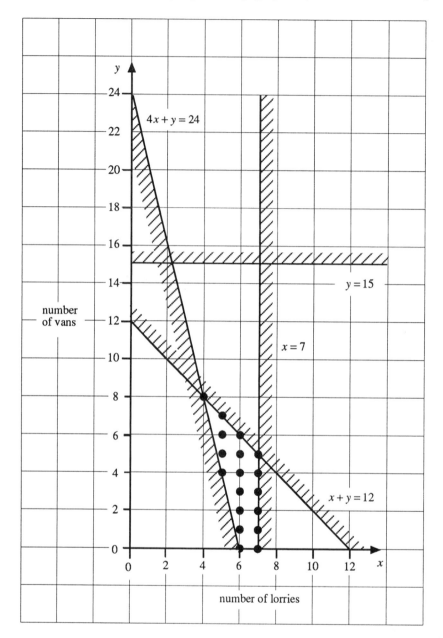

Fig. 7.4

When the four inequalities are shown on the graph together (Fig. 7.4), look for the region with no shading. In this area there are 18 arrangements of lorries and vans that are capable of doing the job.

Now what is the cheapest way to deliver the parcels?

Seven lorries must be cheaper than seven lorries plus some vans, so all the points above (7, 0) can be ignored. Similarly all those above (6, 0) and (5, 4) can also be disregarded. Six lorries are clearly cheaper than seven so the only points left to consider are (4, 8), (5, 4) and (6, 0).

The cost at each of these points is given by $c = 75x + 25y$, so

(4, 8) $\Rightarrow$ $4 \times 75 + 8 \times 25 = £500$

(5, 4) $\Rightarrow$ $5 \times 75 + 4 \times 25 = £475$

(6, 0) $\Rightarrow$ $6 \times 75 + 0 \times 25 = £450$.

Six lorries and no vans can do the job most cheaply and the cost will be £450.

The three constraints bound an area known as the **feasible region** as it contains all the feasible solutions. The **optimal solution** should be sought by examining the corners of the feasible region.

(6, 0) is the cheapest

(7, 5) is most expensive and has greatest capacity

(4, 8) uses most vans.

If a corner has non-integral coordinates (e.g. 10.1 lorries and 1.9 vans), then points near the vertex should be examined. The one closest to the corner is not necessarily the best.

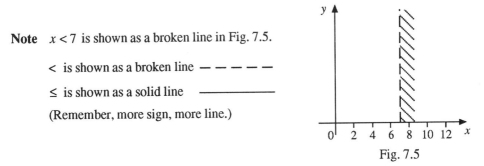

Note $x < 7$ is shown as a broken line in Fig. 7.5.

< is shown as a broken line $--- - - -$

$\leq$ is shown as a solid line $\underline{\hspace{3cm}}$

(Remember, more sign, more line.)

Fig. 7.5

Example

A supermarket has 20 checkouts which are staffed by experienced and trainee till operators. An experienced operator can serve an average of 15 customers per hour and a trainee is expected to serve an average of 10. The store's policy to have a serving capacity of 210 customers per hour can be exceeded but not fallen short of. Union regulations dictate that there must be fewer trainees than experienced staff on the tills at any time. Management policy is that the ratio of trainees to experienced operators should be at least 1 to 3, to keep costs down.

If experienced till operators are paid £7.50 per hour and trainees £4 per hour, what is the cheapest arrangement?

Solution

| **variables:** | x = number of experienced workers |
| | y = number of trainees |

objective function: $P = 7.5x + 4y$ to be minimised

constraints: $15x + 10y \geq 210$ serving rule

$$\Rightarrow \quad 3x + 2y \geq 42$$

$$x > y \qquad \qquad \text{union}$$

$$x + y \leq 20 \qquad \text{number of tills}$$

$$3y \geq x \qquad \qquad \text{management policy}$$

The last of these may cause difficulty. Write down some examples of permitted arrangements that are 'on the border', for example,

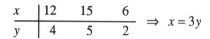

x	12	15	6
y	4	5	2

$\Rightarrow x = 3y$

Then consider how to complete the inequality. The values $x = 6$ and $y = 4$ are satisfactory, giving $x \leq 3y$.

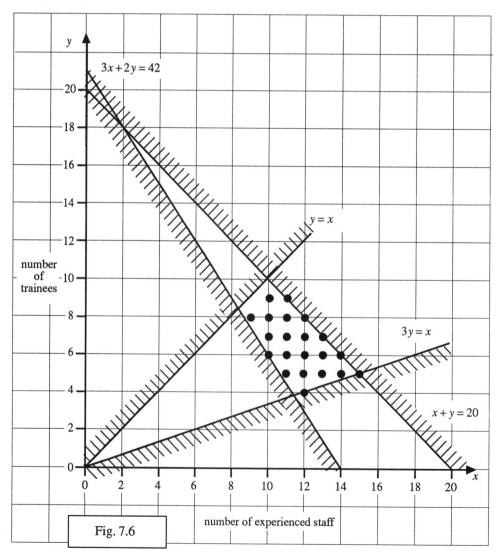

Fig. 7.6

Draw the lines: $x + y = 20$ (0, 20) (20, 0) (10, 10)

$3x + 2y = 42$ (0, 21) (14, 0) (8, 9)

$x = y$ (0, 0) (10, 10) (20, 20)

$x = 3y$ (0, 0) (12, 4) (21, 7)

Check $P = 7.5x + 4y$ near the corners of the unshaded region.

$(9, 8)$ $\Rightarrow$ $7.5 \times 9 + 4 \times 8 = 99.50$

$(11, 5)$ $\Rightarrow$ $7.5 \times 11 + 4 \times 5 = 102.50$

$(10, 6)$ $\Rightarrow$ $7.5 \times 10 + 4 \times 6 = 99$ is cheapest

$(12, 4)$ $\Rightarrow$ $7.5 \times 12 + 4 \times 4 = 106$

The other points can be discarded as before.

Exercise 2.1   A profit of £245 has been made from a village fete, after taking into account costs and donations. A river outing is to be arranged for the helpers. Eighty people would like to go and fifteen of them are prepared to drive the motor boats. The boat-hire firm being used has twelve five-seaters and five eight-seaters available. If the former cost £15 each and the latter £20 each, how can the trip be organised most cheaply?

If three more of the helpers find they can go on the outing but one of the potential drivers has to drop out, can the trip still go ahead, if none of the three will drive?

(Write down inequalities based on: cost, number of drivers, number of people on the trip.)

Exercise 2.2   A car hire firm closes for one week in January because of snow. Eight cars are in the garage and can either be given a service, completely overhauled or left alone. A service costs £40 and a complete overhaul costs £80. If the owner can afford only £520 for the work, what is the greatest number of cars he can have completely overhauled if

(a) the rest are serviced

(b) some are left untouched?

(Write down the inequalities based on: cost, number of cars.)

A complete overhaul takes nine hours of mechanics' time, a service takes three hours, and at least 36 hours of mechanics' time must be utilised to occupy those in the garage. If all the cars are worked on, what is the cheapest the work can be done for?

Exercise 2.3   After some roadworks have been completed, an area of 3000 m² is to be landscaped. The plan involves planting trees, each one costing £20 and requiring space of 30 m², and shrubs, each of which costs £6 and needs 4 m² of space. At least 75 shrubs must be planted to comply the local council regulations. £2400 is available to be spent. Write down three inequalities that must be satisfied, and show them graphically.

If each tree is considered to be five times as beneficial as a shrub for attracting wildlife back to the area, what combination of trees and shrubs should be planted to attract the most wildlife possible?

Exercise 2.4   A wargamer collects Napoleonic soldiers which are available in two sets. Set A contains 20 cavalry and 40 infantry; set B contains 24 cavalry and 16 infantry. If x of set A and y of set B are bought, write down the numbers of cavalry and infantry they provide.

     The collector needs at least 1200 troops, with more infantry than cavalry. Write down two inequalities that must be satisfied and show them graphically.

     If set A costs £3.50 and set B costs £2.20, what is the cheapest way for him to buy the collection he wants?

7.3   DRAWING THE OBJECTIVE FUNCTION
Example

A firm makes dog biscuits and sells them in 1 kg and 0.5 kg bags. The machinery involved can produce up to 5000 kg of biscuits per day and the packaging department can handle at most 8500 packets per day. Market research indicates that no more than 8000 of the 0.5 kg bags will be required by shops per day and the demand for 1 kg bags will not exceed 4000. At least 3500 bags need to be produced to keep the workforce occupied. If there is a profit of 10p on each 1 kg bag and 4p on each of the 0.5 kg size, how should production be planned to maximise the overall profit?

     Should the firm invest in increasing its packaging potential or in buying more machinery to produce the biscuits?

Solution

variables:	x = number of 0.5 kg bags	
	y = number of 1 kg bags	
constraints:	market research	$x \leq 8000$
		$y \leq 4000$
	machines	$\frac{1}{2}x + y \leq 5000$
		$x + 2y \leq 10\ 000$
	packaging	$x + y \leq 8500$
	workforce	$x + y \geq 3500$

$x + 2y = 10\ 000$	$(10\ 000, 0)\ \ (0, 5000)\ \ (4000, 3000)$
$x + y = 8500$	$(8500, 0)\ \ (0, 8500)\ \ (3500, 4000)$
$x + y = 3500$	$(3500, 0)\ \ (0, 3500)\ \ (1500, 2000)$

Objective function: $P = 4x + 10y$ to be maximised.

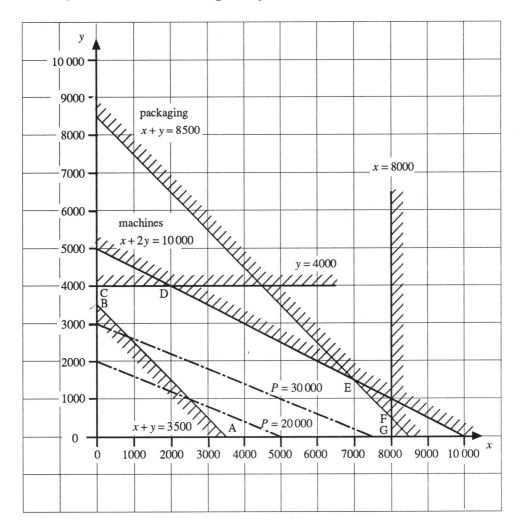

Fig. 7.7

There are seven vertices on the border of the feasible region of which two, A and B, can be disregarded, as G and C must generate more profit. Now C and G can be discounted by comparison with D and F, leaving three to consider.

Rather than work out the value of P at each vertex, or points near it, it is possible to decide which point is optimal by drawing lines of equal profit. This is particularly helpful when there are many vertices to consider.

$$P = 4x + 10y$$

Choose a value for P which is divisible by both 4 and 10, for example, $P = 20\,000$

$$\Rightarrow \quad 20\,000 = 4x + 10y$$

This is a line which passes through (5000, 0) (0, 2000). Now the line can be drawn representing all combinations which produce a profit of 20 000 pence.

$P = 30\,000 \Rightarrow 30\,000 = 4x + 10y$ gives a line parallel to the first which passes through (7500, 0) and (0, 3000) and joins all points that give a profit equal to 30 000 pence.

Other values of P produce further parallel lines and as P increases they move up the graph. Place a ruler on one of the parallel lines drawn and slide it up the page keeping it parallel with the lines. At which point do you leave the feasible region?

The point D is the last one before the ruler leaves the region and the value of P at (2000, 4000) is $4 \times 2000 + 10 \times 4000 = 48\,000$ pence, which is optimal.

Draw lines to represent the objective function on the graphs produced in Exercises 2.1, 2.2, 2.3 and 2.4.

Exercise 3.1   A factory produces cricket bats and tennis rackets. A cricket bat takes one hour of machine time and three hours of craftsman's time, while a tennis racket takes two hours of machine time and one hour of craftsman's time. In one day, the factory has available no more than 56 hours of machine time and 48 hours of craftsman's time.

If the profits on a bat and on a racket are £10 and £5 respectively, find the maximum possible profit to the factory for one day's work.

Exercise 3.2 A market gardener intends to grow carrots and potatoes in a 10 hectare field. The relevant details are

	carrots	potatoes
cost per hectare	£200	£120
labour per hectare	24	8
profit per hectare	£300	£120

If £1480 and 156 man days are available, how should the land be allocated to maximise the profit?

Exercise 3.3   A paper firm takes rolls 36 inches wide and cuts and distributes them to customers. There are currently orders for 600 feet of paper 10 inches wide and 400 feet of paper of width 6 inches, to be satisfied. The firm's machines can cut a roll into at most 5 strips.

The widths required can be produced by using plan A or plan B, illustrated in Fig. 7.8.

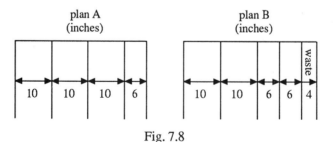

Fig. 7.8

If x feet are cut according to plan A and y feet according to plan B, explain why
$$3x + 2y \geq 600$$
Write down a further inequality that must be satisfied (other than $x \geq 0$, $y \geq 0$).
How should the cutting be organised so as to use the minimum length of roll?

Exercise 3.4   The robot T4-2 can walk at 1.5 m s⁻¹ or run at 4 m s⁻¹. When walking it consumes power at one unit per metre and at three times this rate while running. If its batteries are fully charged to 3000 units, what is the greatest distance it can cover in ten minutes?

(Hint: let x = distance run in metres, y = distance walked in metres and write down two inequalities based on power and time.)

7.4   NOTES

Graphical methods may also be applied to situations with non-linear constraints or objective function.

Example

A firm plans to manufacture rectangular rugs. The area of each rug must be at least 12 ft². The cost of stitching the sides is 10p/ft and of decorating the ends is 20p/ft. If no more than £1.20 is to be spent on the sides and the ends, and the sides are at least as long as the ends, what is   (a) the maximum possible perimeter   (b) the largest feasible rug?

Solution

Let x be the length of end and y be the length of side

area: $xy \geq 12$ cost: $20x + 10y \leq 120$

$$\Rightarrow 2x + y \leq 12$$

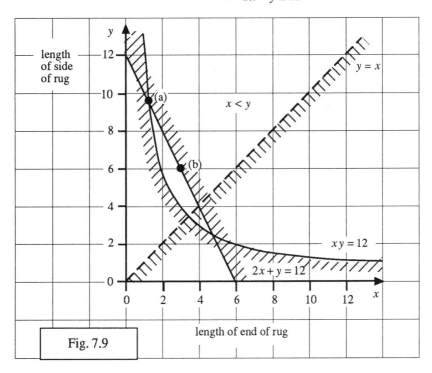

Fig. 7.9

The graph shows the inequalities, and the solutions for (a) and (b) are indicated.

8

Linear programming: simplex method

8.1 INVESTIGATION

A college committee has £2700 to spend on furniture for a common room. They decide to buy two types of easy chairs and some coffee tables. Each easy chair with arms costs £20 and occupies 0.4 m² of floor space, while the basic chair is £5 cheaper and takes up 0.35 m²; each coffee table costs £10 and occupies 1 m². Due to safety regulations only 69.9 m² is available for furniture. It has been decided that at most, one-third of the chairs will be of the basic type. How should the money be spent to buy as many pieces of furniture as possible?

8.2 SIMPLEX METHOD

The simplex method starts at a vertex of the feasible region where the value of the objective function is calculated. Routes from here are compared and, of those which improve the value, the one that does so most quickly is chosen. This is like the problem of finding the best route from the bottom to the top of the diagram in Fig. 8.1.

One method is to look at those ladders that go upwards and always select the one that does so most steeply. Since the lengths of the ladders are also important, the method may not produce the route with the fewest intermediate stages.

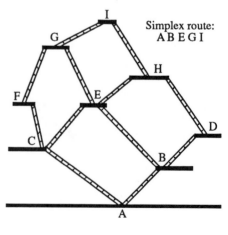

Simplex route:
A B E G I

Fig. 8.1

Example

The robot T4-2 can walk at a rate of 1.5 m s^{-1} or run at 4 m s^{-1}. When walking it consumes power at one unit per metre and at three times this rate while running. If its batteries are fully charged to 3000 units, what is the greatest distance it can cover in ten minutes?

Solution

Let x be the distance run in metres and y be the distance walked in metres, then

time: $\qquad\qquad\qquad \dfrac{x}{4} + \dfrac{y}{1.5} \leq 600 \qquad$ in seconds

$\times 12 \quad \Rightarrow \qquad 3x + 8y \leq 7200$

power: $\qquad\qquad\qquad 3x + \dfrac{y}{1} \leq 3000$

objective function: $\quad x + y = P$ to be maximised.

Introduce **slack variables** u and v

$$3x + 8y + u \qquad = 7200$$

$$3x + y \qquad + v = 3000$$

to 'take up the slack' in the constraints, turning the inequalities into equations. u and v have real meanings in terms of the example, with u being twelve times the time not used (i.e. less than ten minutes) and v being the power remaining in the batteries. If slack variables had been introduced to the original inequalities they would have simply represented the unused time and power.

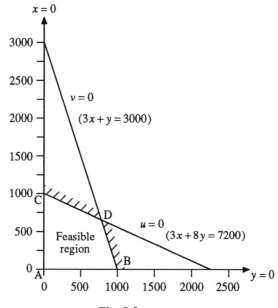

Fig. 8.2

The feasible region is bounded by the four lines

$$x = 0, \quad y = 0,$$
$$3x + 8y + u \quad\quad = 7200$$
$$3x + y \quad\quad +v = 3000$$

and by plotting $3x + 8y = 7200$ for one and $3x + y = 3000$ for another, as before. The lines can also be described as

$$x = 0, \quad y = 0,$$
$$u = 0, \quad v = 0.$$

A basic feasible solution that satisfies the constraints

$$3x + 8y + u \quad\quad = 7200$$
$$3x + y \quad\quad +v = 3000$$

is $x = 0$, $y = 0$, $u = 7200$, $v = 3000$. u and v are called the **basic variables** as they form the basis for the solution suggested. This solution was easy to find as the constraints are in **canonical form**, that is, each is the sole container of a variable that appears exactly once. However, the solution is hardly likely to be optimal since $x = 0$, $y = 0 \Rightarrow P = 0$. The simplex method starts then, at $x = 0$, $y = 0$, (i.e. A) and looks to improve P.

Increasing either x or y will improve the value of P. If you had, for example, $P = 3x + 2y$ then it would seem sensible to start by increasing x rather than y as this would have the greater effect on P (the steeper ladder). However, it might be possible to increase y considerably more than x (the longer ladder) but it is simpler to take the x and proceed as it is going in the right direction and may well be the quickest route to optimality.

With $P = x + y$, either x or y may be chosen to be increased, so select one, x say. The problem is to decide just how much it can be changed. Since x, y, u, and v are all non-negative variables, consideration of the constraints.

(1) $3x + 8y + u \quad\quad = 7200$ $\Rightarrow$ 2400

(2) $3x + y \quad\quad +v = 3000$ $\Rightarrow$ 1000

shows that x can be increased to 2400 in the first constraint, but only to 1000 in the second (remember $y = 0$). When $x = 1000$, v becomes zero, so x and u form the basis of the new solution while $v = 0$ and $y = 0$. Now you have moved to B whereas increasing y would have brought you to C.

Now put the constraints in a new canonical form based on x and u,

(1) – (2) $\Rightarrow$ $7y + u - v = 4200$ (3)

(2) ÷ 3 $\Rightarrow$ $x + \dfrac{y}{3} + \dfrac{v}{3} = 1000$ (4)

and write the objective function

$$x + y = P$$

in terms of the non-basic variables y and v.

Subtracting   (4) $\Rightarrow$ $y - \dfrac{y}{3} - \dfrac{v}{3}$ $= P - 1000$

$\Rightarrow$ $\dfrac{2y}{3} - \dfrac{v}{3}$ $= P - 1000$

Now $P = 1000$ at the vertex B where $y = 0$ and $v = 0$. Increasing the value of y will improve P so again look at the constraints

(3) $7y + u - v = 4200$ $\Rightarrow 600$

(4) $x + \dfrac{y}{3} + \dfrac{v}{3} = 1000$ $\Rightarrow 3000$

which show that y cannot increase beyond 600 without u becoming negative (remember $v = 0$). y will increase to 600 and u will be reduced to zero and leave the basis. $u = 0$, $v = 0$ is the point D.

Next put the constraints in canonical form based on x and y.

(3) ÷ 7 $\Rightarrow$ $y + \dfrac{u}{7} - \dfrac{v}{7} = 600$ (5)

(4) $- \dfrac{1}{3}(5)$ $\Rightarrow$ $x - \dfrac{u}{21} + \dfrac{8v}{21} = 800$ (6)

and write P in terms of the non-basic variable x and y.

Subtracting $\dfrac{2}{3}(5)$ $\Rightarrow$ $-\dfrac{2}{3}\left(-\dfrac{u}{7} + \dfrac{v}{7}\right) - \dfrac{v}{3}$ $= P - 1400$

$\Rightarrow$ $-\dfrac{2u}{21} - \dfrac{5v}{21}$ $= P - 1400$

Since $u = 0$ and $v = 0$, P has the value 1400 at D, and since increasing either u or v will lower the value, optimality has been achieved.

T4-2 can travel 1400 m of which 800 are run and 600 walked.

The method may be summarised as

   (1)   introduce slack variables

   (2)   find a basic feasible solution

   (3)   obtain canonical form

   (4)   express the objective function in terms of the non-basic variables

   (5)   find the variable to enter the basis

   (6)   find the variable to leave the basis

   (7)   repeat steps 3 to 6 until no change at step 5 will improve the value of P.

8.3   THE SIMPLEX TABLEAU

The working of the simplex method may be neatly displayed in tabular form. For the example in section 8.2,

constraints:
$$3x + 8y \leq 7200$$
$$3x + y \leq 3000$$

introduce slack variables
$$3x + 8y + u = 7200$$
$$3x + y + v = 3000$$

maximise
$$x + y = P$$

basic feasible solution
$$x = 0, y = 0, u = 7200, v = 3000$$

Note Step 6 is done by dividing the 'value' by 'the element corresponding to the variable to enter the basis' for each row. For the first iteration, $7200 \div 3 = 2400$ and $3000 \div 3 = 1000$ so v leaves the basis, as it becomes negative before u does, i.e. when $x = 1000$.

	x	y	u	v	value	basis	check
(1)	3	8	1	•	7200	u	7212
(2)	3*	1	•	1	3000	v	3005
(3)	1	1	•	•	0	P	2
	↑						
(4)	•	7*	1	−1	4200	u	4207
(5)	1	$\frac{1}{3}$	•	$\frac{1}{3}$	1000	$x^\wedge$	$1001\frac{2}{3}$
(6)	•	$\frac{2}{3}$	•	$-\frac{1}{3}$	−1000	P	$1000\frac{1}{3}$
		↑					
(7)	•	1	$\frac{1}{7}$	$-\frac{1}{7}$	600	y	601
(8)	1	•	$-\frac{1}{21}$	$\frac{8}{21}$	800	x	$801\frac{1}{3}$
(9)	•	•	$-\frac{2}{21}$	$-\frac{5}{21}$	−1400	P	$-1400\frac{1}{3}$

The use of fractions avoids long strings of decimals, $\dfrac{1}{7} = 0.\dot{1}4285\dot{7}$.

↑ indicates the variable to enter the basis.

* indicates, by row, the variable to leave the basis.

The row containing * is called the **pivot row** and is used for adjusting the other rows to produce a new canonical form.

Check: the column on the right is initially the sums of the elements of the rows. In succeeding tableaux these are operated on in the same way as the other figures in the rows and if all has been done correctly they should continue to be row sums.

$$(5) = (2) \div 3 \qquad (4) = (1) - (2) \qquad (6) = (3) - (5)$$

$$(7) = (4) \div 7 \qquad (8) = (5) - (7) \div 3 \qquad (9) = (6) - \frac{2}{3}(7)$$

Now,
$$-\frac{2u}{21} - \frac{5v}{21} = -1400 + P$$

$$\Rightarrow \quad P = 1400 \text{ when } u = 0, v = 0, x = 800, y = 600$$

Example

Find the maximum value of $x + 3y = P$ subject to the constraints

$$x + y \ \leq 12$$
$$x + 2y \leq 14$$
$$x + 4y \leq 24$$

Solution

Step 1   Introduce slack variables

(1)	$x + y \ + \ u$			$= 12$
(2)	$x + 2y$	$+v$		$= 14$
(3)	$x + 4y$		$+ w$	$= 24$

Step 2   Find a basic feasible solution

$x = 0, y = 0, u = 12, v = 14, w = 24$

Step 3   Obtain a canonical form.
This already exists as the basic variables u, v and w occur once each in different equations with coefficient of one.

Step 4   Express the objective function in terms of the non-basic variables. The non-basic variables are x and y and the objective function P is given in terms of them.

(4) $x + 3y \ \ = \ \ P$

The initial tableau is

	x	y	u	v	w	value	basis	check
(1)	1	1	1	•	•	12	u	15
(2)	1	2	•	1	•	14	v	18
(3)	1	4	•	•	1	24	w	30
(4)	1	3	•	•	•	0	P	4

Step 5   Find the variable to enter the basis.

This is identified by inspecting the row of the objective function and selecting the variable with the largest positive number in it. In this case y has the greatest value in the row for P, that is, 3 and it is indicated by an arrow $\uparrow$.

Step 6   Find the variable to leave the basis.

y is to enter the basis and the one it replaces is found by dividing each of the values by the corresponding number in the y column and selecting the lowest positive one. (If there are several lowest answers to the divisions, any can be used). Here w leaves the basis since when y reaches 6, w becomes zero.

	x	y	u	v	w	value	basis	check	
(1)	1	1	1	•	•	12	u	15	$12 \div 1 = 12$
(2)	1	2	•	1	•	14	v	18	$14 \div 2 = 7$
(3)	1	4*	•	•	1	24	w	30	$24 \div 4 = 6$
(4)	1	3	•	•	•	0	P	4	

$\uparrow$

y will enter the basis and, in the new canonical form, there will be a 1 in the position marked * and any other numbers in this column must be eliminated.

The element marked with a * is called the **pivot element** and its row is known as the **pivot row**. It is this row that is used to produce the canonical form.

Step 3   Obtain a canonical form.

You now need a new canonical form with a basis consisting of u, v and y. This will have the form that follows.

	x	y	u	v	w	value	basis	check
(5)		•	1	•			u	
(6)		•	•	1			v	
(7)		1	•	•			y	
(8)		•	•	•			P	

The aim then is to produce a 1 in the position indicated by the pivot element and to use the pivot row to eliminate all the other numbers in its column.

	x	y	u	v	w	value	basis	check
$(1)-(3) \div 4 = (5)$	$\frac{3}{4}$	•	1	•	$-\frac{1}{4}$	6	u	$7\frac{1}{2}$
$(2)-(3) \div 2 = (6)$	$\frac{1}{2}$	•	•	1	$-\frac{1}{2}$	2	v	3
$(3) \div 4 = (7)$	$\frac{1}{4}$	1	•	•	$\frac{1}{4}$	6	y	$7\frac{1}{2}$
$(4)-\frac{3}{4} \times (3) = (8)$	$\frac{1}{4}$	•	•	•	$-\frac{3}{4}$	-18	P	$-18\frac{1}{2}$

The same operation has been done to each element in a row and if all has been completed correctly the check column will continue to be sums of the elements in the rows.

Step 4   Express the objective function in terms of the non-basic variables (x and w here).

This has already been done by placing dots in the desired places in row 8. Dots are used to signify positions which will be kept empty of any numbers. Zeroes may occur elsewhere during the working and would be shown as such.

Step 5   Find the variable to enter the basis. The objective function has only one positive entry and its position indicates that x will enter the basis. (↑)

Step 6   Find the variable to leave the basis.

	x	y	u	v	w	value	basis	check	
(5)	$\frac{3}{4}$	•	1	•	$-\frac{1}{4}$	6	u	$7\frac{1}{2}$	$6 \div \frac{3}{4} = 8$
(6)	$\frac{1}{2}$ *	•	•	1	$-\frac{1}{2}$	2	v	3	$2 \div \frac{1}{2} = 4$
(7)	$\frac{1}{4}$	1	•	•	$\frac{1}{4}$	6	y	$7\frac{1}{2}$	$6 \div \frac{1}{4} = 24$
(8)	$\frac{1}{4}$	•	•	•	$-\frac{3}{4}$	-18	P	$-18\frac{1}{2}$	

↑

The smallest positive answer is 4, in v's row, so it is v that will leave the basis.

Step 3   Obtain the canonical form.

Step 4   Express the objective function in terms of the non-basic variables. A tableau like the one that follows is required.

	x	y	u	v	w	value	basis	check
(9)	•	•	1				u	
(10)	1	•	•				x	
(11)	•	1	•				y	
(12)	•	•	•				P	

A one is needed to replace the pivot element and the pivot row is used to eliminate the numbers from the x column.

	x	y	u	v	w	value	basis	check
$(5) - \frac{3}{2}(6) = (9)$	•	•	1	$-\frac{3}{2}$	$\frac{1}{2}$	3	u	3
$(6) \times 2 = (10)$	1	•	•	2	-1	4	x	6
$(7) - \frac{1}{2}(6) = (11)$	•	1	•	$-\frac{1}{2}$	$\frac{1}{2}$	5	y	6
$(8) - \frac{1}{2}(6) = (12)$	•	•	•	$-\frac{1}{2}$	$-\frac{1}{2}$	-19	P	-20

Step 5   Find the variable to enter the basis.

As the objective function has no positive entries there is no candidate for inclusion in the basis. The objective function's row is equivalent to

$$-\frac{1}{2}v-\frac{1}{2}w=-19+P$$
$$\Rightarrow \quad 19-\frac{1}{2}v-\frac{1}{2}w=P$$

The maximum value is 19 when $v=0$ and $w=0$, which they currently do.

So $P_{max}=19$ when $x=4, y=5$.

The whole process can be written as follows:

	x	y	u	v	w	value	basis	check	
(1)	1	1	1	•	•	12	u	15	$12 \div 1 = 12$
(2)	1	2	•	1	•	14	v	18	$14 \div 2 = 7$
(3)	1	4*	•	•	1	24	w	30	$24 \div 4 = 6$
(4)	1	3	•	•	•	0	P	4	
		↑							
(1)−(3)÷4 = (5)	$\frac{3}{4}$	•	1	•	$-\frac{1}{4}$	6	u	$7\frac{1}{2}$	$6 + \frac{3}{4} = 8$
(2)−(3)÷2 = (6)	$\frac{1}{2}$*	•	•	1	$-\frac{1}{2}$	2	v	3	$2 + \frac{1}{2} = 4$
(3)÷4 = (7)	$\frac{1}{4}$	1	•	•	$\frac{1}{4}$	6	y	$7\frac{1}{2}$	$6 + \frac{1}{4} = 24$
(4)−$\frac{3}{4}$(3) = (8)	$\frac{1}{4}$	•	•	•	$-\frac{3}{4}$	− 18	P	$-18\frac{1}{2}$	
	↑								
(5)−$\frac{3}{2}$(6) = (9)	•	•	1	$-\frac{3}{2}$	$\frac{1}{2}$	3	u	3	
(6)×2 = (10)	1	•	•	2	− 1	4	x	6	
(7)−$\frac{1}{2}$(6) = (11)	•	1	•	$-\frac{1}{2}$	$\frac{1}{2}$	5	y	6	
(8)−$\frac{1}{2}$(6) = (12)	•	•	•	$-\frac{1}{2}$	$-\frac{1}{2}$	− 19	P	− 20	

$$\Rightarrow \quad P_{max} = 19 \text{ when } x = 4, y = 5$$

Exercise 3.1   Repeat the example from section 8.2, bringing y into the basis at the first iteration.

Exercise 3.2   Maximise: $P = 2x + y$

subject to: $4x + 3y \le 300$

$2x + 9y \le 360$

$x \ge 0, y \ge 0$

Exercise 3.3   Maximise: $P = x + y$

subject to: $x + 2y \le 80$

$3x + 2y \le 120$

$x \ge 0, y \ge 0$

Exercise 3.4   Maximise: $P = x + 2y$

subject to: $3x + 4y \le 1700$

$2x + 5y \le 1600$

$x \ge 0, y \ge 0$

Exercise 3.5   Maximise: $P = 4x + 5y$

subject to: $5x + 2y \le 30$

$5x + 7y \le 35$

$2x + 5y \le 20$

$x \ge 0, y \ge$

Exercise 3.6   A factory makes cricket bats and tennis rackets. A bat takes one hour of machine time and three hours of craftman's time, while a racket takes two hours of machine time and one hour of craftman's time. In a day the factory has available no more than 56 hours of machine time and 48 hours of craftman's time. If the profits on a bat and a racket are £10 and £5 respectively, find the maximum possible profit.

Exercise 3.7   A market gardener intends to split a 10 hectare field between carrots and potatoes. The relevant details are

	carrots	potatoes	
cost per hectare	£200	£120	
labour per hectare	24	8	(man-days)
profit per hectare	£300	£120	

If £1480 and 156 man-days are available, how should the land be allocated to maximise the profit?

8.4   ≥ CONSTRAINTS
Example

Maximise: $4x + 10y = P$

subject to: $x + y \quad \le 8500$

$x + 2y \le 10000$

$x + y \quad \ge 3500$

(See Chapter 7, *Linear programming: graphical*, Example in section 7.3.)

Solution
Introduce slack variables

$$x + y\ + u \qquad\qquad = 8500$$
$$x + 2y \qquad + v \qquad = 10000$$
$$x + y \qquad\quad -w\ = 3500$$

Now it is not simple to see a basic feasible solution since $x = 0$, $y = 0$, $u = 8500$, $v = 10000$, $w = -3500$ is not acceptable in view of the non-negativity requrement.

One method of producing a basic feasible solution is to use two slack variables in the last constraint, so

$$x + y + u \qquad\qquad = 8500$$
$$x + 2y\ + v \qquad\quad = 10000$$
$$x + y \qquad + z - w\ = 3500$$
$$4x + 10y \qquad = P$$

which gives a basic feasible solution $x = 0$, $y = 0$, $w = 0$, $u = 8500$, $v = 10000$, $z = 3500$.

The auxilary slack variable is part of the basis and the aim now is to remove z from this set. This can be done using row (3) to bring x or y into the basis in its place.

	x	y	u	v	z	w	value	basis	check
(1)	1	1	1	•	•	0	8500	u	8503
(2)	1	2	•	1	•	0	10000	v	10004
(3)	1	1*	•	•	1	−1	3500	z	3502
(4)	4	10	•	•	•	0	0	P	14
		↑							
(1) − (3) = (5)	0	•	1	•	−1	1	5000	u	5001
(2) − 2(3) = (6)	−1	•	•	1	−2	2	3000	v	3000
(3) = (7)	1	1	•	•	1	−1	3500	y	3502
(4) − 10(3) = (8)	−6	•	•	•	−10	10	−35000	P	−35006

Now delete the z column so it never returns to the basis, and continue. Since z is currently zero its disappearance does not affect the values of the other variables.

8.5   THREE DIMENSIONS
Example
Look again at the example given in section 8.1.

Solution
Let x = number of chairs with arms, y = number of basic chairs and z = number of tables.

space		$0.4x + 0.35y + z \leq 69.9$
	$\Rightarrow$	$8x + 7y + 20z \quad \leq 1398$

cost		$20x + 15y + 10z \leq 2700$
	$\Rightarrow$	$4x + 3y + 2z \quad \leq 540$

'look'		$y \leq \frac{1}{3}(x + y)$
	$\Rightarrow$	$3y - (x + y) \leq 0$
	$\Rightarrow$	$-x + 2y \quad \leq 0$

Maximise: $x + y + z = P$

Introduce slack variables

$$8x + 7y + 20z + u \quad\quad\quad = 1398$$
$$4x + 3y + 2z \quad\quad +v \quad\quad = 540$$
$$-x + 2y \quad\quad\quad\quad +w \;=\; 0$$

Basic feasible solution: $x = 0; \quad y = 0; \quad z = 0; \quad u = 1398; \quad v = 540; \quad w = 0.$

Notice that the number of elements in the basic feasible solution, and hence the final optimal solution, is determined by the number of constraints and not the number of variables in the problem. It is quite possible that a basic variable has zero as its value, as here with $w = 0$.

	x	y	z	u	v	w	value	basis	check
(1)	8	7	20	1	•	•	1398	u	1434
(2)	4*	3	2	•	1	•	540	v	550
(3)	−1	2	0	•	•	1	0	w	2
(4)	1	1	1	•	•	•	0	P	3
	↑								
(5)	•	1	16*	1	−2	•	318	u	334
(6)	1	$\frac{3}{4}$	$\frac{1}{2}$	•	$\frac{1}{4}$	•	135	x	$137\frac{1}{2}$
(7)	•	$2\frac{3}{4}$	$\frac{1}{2}$	•	$\frac{1}{4}$	1	135	w	$139\frac{1}{2}$
(8)	•	$\frac{1}{4}$	$\frac{1}{2}$	•	$-\frac{1}{4}$	•	−135	P	$-134\frac{1}{2}$
			↑						

(9)	$\bullet$	$\frac{1}{16}$	1	$\frac{1}{16}$	$-\frac{1}{8}$	$\bullet$	$19\frac{7}{8}$	z	$20\frac{7}{8}$
(10)	1	$\frac{23}{32}$	$\bullet$	$-\frac{1}{32}$	$\frac{5}{16}$	$\bullet$	$125\frac{1}{16}$	x	$127\frac{1}{16}$
(11)	$\bullet$	$2\frac{23}{32}*$	$\bullet$	$-\frac{1}{32}$	$\frac{5}{16}$	1	$125\frac{1}{16}$	w	$129\frac{1}{16}$
(12)	$\bullet$	$\frac{7}{32}$	$\bullet$	$-\frac{1}{32}$	$-\frac{3}{16}$	$\bullet$	$-144\frac{15}{16}$	P	$-144\frac{15}{16}$
		$\uparrow$							
(13)	$\bullet$	$\bullet$	1	$\frac{11}{174}$	$-\frac{23}{174}$	$-\frac{2}{87}$	17	z	$17\frac{79}{87}$
(14)	1	$\bullet$	$\bullet$	$-\frac{2}{87}$	$\frac{20}{87}$	$-\frac{23}{87}$	92	x	$92\frac{82}{87}$
(15)	$\bullet$	1	$\bullet$	$-\frac{1}{87}$	$\frac{10}{87}$	$\frac{32}{87}$	46	y	$47\frac{41}{87}$
(16)	$\bullet$	$\bullet$	$\bullet$	$-\frac{5}{174}$	$-\frac{37}{174}$	$-\frac{7}{87}$	-155	P	$-155\frac{28}{87}$

$(6) = (2) \div 4$ $(5) = (1) - 2(2)$ $(7) = (3) + (6)$ $(8) = (4) - (6)$

$(9) = (5) \div 16$ $(10) = (6) - (5) \div 32$ $(11) = (7) - (5) \div 32$ $(12) = (8) - (5) \div 32$

$(15) = (11) \div 2\frac{23}{32}$ $(13) = (9) - (15) \div 16$ $(14) = (10) - \frac{23}{87}(11)$ $(16) = (12) - \frac{7}{87}(11)$

$\Rightarrow \quad P = 155$ is the best total.

While graphical methods work well for two variable problems, the simplex method can be used in cases that involve any number of variables.

The simplex method moves around the feasible region from vertex to vertex so it fails when the optimum solution is not at such a point. This will happen when only integer values are admissible for the variables, as in the example here, and the vertices have non-integral coordinates.

8.6 MINIMISATION PROBLEMS
Example

$$\text{Minimise:} \quad x + y = P \tag{1}$$
$$\text{subject to:} \quad 3x + 2y \geq 600$$
$$x + 2y \geq 400$$

(See Chapter 7, *Linear programming: graphical*, Exercise 3.3.)

Solution
Introduce slack variables

$$3x + 2y - u \qquad\quad = 600 \tag{2}$$
$$x + 2y \qquad -v = 400 \tag{3}$$

Basic feasible solution (can use method of section 8.4)

e.g. ; $x = 400;$ $y = 0;$ $v = 0;$ $u = 600$

Obtain canonical form

$$3(3) - (2) \qquad 4y + u - 3v = 600 \tag{4}$$
$$(3) \qquad x + 2y \quad - v = 400 \tag{5}$$

Express the objective function in terms of the non-basic variables y and v.

$$(1)-(3) \qquad -y \qquad +v = P - 400$$

Now the problem is to minimise P, which is the same as maximising $-P$,

i.e.   maximise $y - v = -P + 400$

x	y	u	v	value	basis	check
$\bullet$	4	1	-3	600	u	602
1	2	$\bullet$	-1	400	x	402
$\bullet$	1	$\bullet$	-1	400	$-P$	400

               ↑   etc.

Exercise 6.1   Maximise: $x + 2y = P$

subject to: $-x + 3y \leq 10$

$x + y \leq 6$

$x \geq 0, y \geq 0$

Exercise 6.2   Maximise: $8x + 19y + 7z = P$

subject to: $x + 3y + 3z \leq 18$

$3x + 4y + z \leq 9$

$x \geq 0, y \geq 0, z \geq 0$

Exercise 6.3   Maximise: $x = P$

subject to: $x + y \leq 30$

$x - 2y \leq 0$

$2x + y \geq 30$

$x \geq 0, y \geq 0$

Exercise 6.4   A woman takes over an 18 hectare smallholding on which are grown three crops, A, B and C. She wants to find the best arrangement to maximise her profit from the crops. Use the details below to plan her land use to best advantage.

	Crop A	Crop B	Crop C
cost per hectare	£50	£30	£10
labour per hectare	16 days	9 days	5 days
profit	£250	£175	£75

Totals available: £650 and 205 days of labour.

Exercise 6.5   T4-3 is an improvement on the earlier T4-2 model. As well as being able to run at 4 m s^{-1} and walk at 1.5 m s^{-1} it can trot at 3 m s^{-1}. When running it uses 3 units of power per metre, walking uses 1 unit per metre and trotting uses 2 units per metre. With batteries fully charged with 3000 units of power, how far can T4-3 travel in 10 minutes?

(Let x = number of metres walked, y = number of metres trotted and z = number of metres run, and formulate constraints by considering power and time.)

Exercise 6.6 A reel of cloth 1 m wide is cut into strips of width 15 cm, 30 cm and 40 cm for sale to customers, by the following methods.

method	1	2	3
no. of 15 cm pieces	0	1	2
no. of 30 cm pieces	2	0	1
no. of 40 cm pieces	1	2	1
waste	–	5 cm	–
length cut	x m	y m	z m

The firm has orders for 160 m of width 15 cm, 140 m of width 30 cm and 200 m of width 40 cm. Explain why $y + 2z \geq 160$ and write down two other constraints (other than $x \geq 0$, $y \geq 0$, $z \geq 0$). How should the reel be cut if the total length used is to be as small as possible?

8.7   NOTES

Minimise: $30x + 35y + 20z = P$

subject to: $5x + 5y + 2z \geq 4$

$2x + 7y + 5z \geq 5$

This can be turned into a maximising problem by the method of section 8.6 and solved by the simplex method. What do you need to do, as there are three variables in the problem?

The solution is

$P = 27.4$ when $x = 0.12$, $y = 0.68$, $z = 0$.

Now the dual problem, Exercise 3.5,

maximise: $4x + 5y = P$

subject to: $5x + 2y \leq 30$

$5x + 7y \leq 35$

$2x + 5y \leq 20$

is solvable graphically and has optimal value

$P = 27.4$ when $x = 5.6$, $y = 1$.

A problem with $n > 2$ variables and two constraints can be transformed into its dual which has only two variables and may, therefore, be solved graphically.

A minimisation may be solved by considering the dual problem which is one of maximising.

The values of the variables may still need to be found.

9

The transportation problem

9.1 INVESTIGATIONS

Investigation 1

A coach hire firm has four coaches at depot D1 and five coaches at depot D2. All nine are to be used on the same day to take parties on excursions. School S has booked four coaches, college C has ordered three and the remaining two will be used by holiday camp H.

If the distances between the depots and various departure points are as given in the table, which depots should supply the different customers to keep the overall mileage to a minimum (and hence the profit to a maximum)?

		S	C	H	
Depots	D1	10	8	9	distances
	D2	11	3	12	in miles

Investigation 2

A general has to embark 5500 personnel for humanitarian operations abroad. The bases B1, B2 and B3 will provide the numbers of people shown in the boxes in Fig. 9.1, and the transport available at the ports P1, P2 and P3 is shown in the circles.

How many personnel should move from the different bases to the various ports?

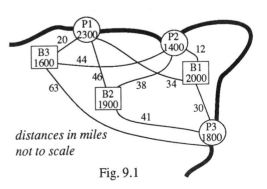

distances in miles
not to scale

Fig. 9.1

9.2 THE TRANSPORTATION ARRAY
Example

Five customers have ordered coal from the central supplier who will arrange distribution from three pits. The customers C1, C2, C3, C4 and C5 require 36 000, 60 000, 24 000, 45 000 and 15 000 tonnes respectively. The pits P1, P2 and P3 can supply 45 000, 60 000 and 75 000 tonnes, in that order. The transportation costs in £'s per tonne are given in the table below.

Customers

		C1	C2	C3	C4	C5
	P1	1	3	10	4	7
Pits	P2	7	10	6	6	3
	P3	4	13	4	5	7

How should the distribution manager arrange to supply the customers most economically? All the information is shown in a single array.

	C1	C2	C3	C4	C5	Supplies
P1	1	3	10	4	7	45
P2	7	10	6	6	3	60
P3	4	13	4	5	7	75
Demands	36	60	24	45	15	180

(demands and supplies in thousands of tonnes)

The aim is to put numbers in some of the available 15 cells in such a way that the entries in row P1 add up to 45, those in column C3 have a sum equal to 24, etc. and the arrangement gives a minimum cost.

This is accomplished by a **trial and improvement** method. In order not to have to go through the improvement process too often an attempt should be made to find an initial feasible solution that is not too many steps away from the optimal arrangement. The **lowest-costs-first method** fills the cells with lowest transportation costs first. In the array drawn up below, cell P1 C1 is selected first and the largest number possible, 36 to satisfy the demand of C1, is placed there. A 'dash' can be put in each of P2 C1 and P3 C1 since C1 will not require supplies from either P2 or P3. This gives a reduced array in which P1 has a supply of 9 available for customers C2, C3, C4 and C5.

	C1	C2	C3	C4	C5	Supplies
P1	36 1	3	10	4	7	45
P2	7	10	6	6	3	60
P3	4	13	4	5	7	75
Demands	36	60	24	45	15	180

Either cell P1 C2 or cell P2 C5 can be chosen next. For example, put 15 in P2 C5 and dashes in P1 C5 and P3 C5. Next select P1 C2 and add 9 there, and dashes in P1 C3 and P1 C4. P3 C3 is the next choice and 24 is put there, and a dash in P2 C3. Finally 45 is put in P3 C4 forcing a dash in P2 C4, 6 in P3 C2 and 45 in P2 C2.

	C1	C2	C3	C4	C5	Supplies
P1	36 ₁	9 ₃	10	4	7	45
P2	— ₇	45 ₁₀	6	6	15 ₃	60
P3	₄	6 ₁₃	24 ₄	45 ₅	7	75
Demands	36	60	24	45	15	180

This distribution satisfies all the customers' requirements, and costs

$36 \times 1 + 9 \times 3 + 45 \times 10 + 6 \times 13 + 24 \times 4 + 45 \times 5 + 15 \times 3 = £957$ (thousand)

– but is it the cheapest arrangement?

The six units in cell P3 C2 is clearly expensive so it would be a good idea to redirect this six to another consumer and to re-supply C2 from a different pit, if possible. This can be done by re-routing the six units from P3 to C1 instead of C2, sending six fewer from P1 to C1 to keep C1 happy, increasing P1's supply to C2 to 15 using the six not now required by C1, and completing C2's order.

This saves $6 \times (13 - 4) = 54$ by moving six from P3 C2 to P3 C1, and costs $6 \times (3 - 1) = 12$ by moving six from P1 C1 to P1 C2. Thus the overall cost can be lowered by £42 (thousand).

This improvement involves four cells. To check **all** the thirty cycles of four cells would take a considerable time and would need to be followed by a look at alterations using more than four positions in the array, so a method for finding improvements is called for.

The transportation costs, given at the bottom right of each cell, can be split into 'part costs' and assigned to the pits and consumers involved. This division between the sender and recipient can be thought of as 'despatch' and 'reception' costs. You can start by assigning a part cost to any row or column, but choosing the one with the largest number of entries is best.

For example, assign P3 a part cost of 0 and then C2 must be given 13; C3 gets 4 and C4 gets 5, if you are to explain the costs you are paying for the routes you have chosen to employ. These part costs are calculated using only those cells which form part of the solution you are seeking to improve. Now P2 C2 forces P2 to be assigned -3 and P1 gets -10. The negatives might seem strange but the part costs assigned to the pits $P1 = -10$, $P2 = -3$ and $P3 = 0$, simply indicate that it should be 10 units cheaper to deal with P1 than to get supplies from P3. These costs are simply relative to each other; negatives arise as a result of assigning 0 to P3. An initial allocation of a part cost of 0 to P1 would have produced P1 : 0, P2 : 7, P3 : 10 with no relative change.

Now P1 C1 gives C1 a part cost of 11 and P2 C5 gives C5 a part cost of 6.

The sums of the part costs are put in the bottom left corners of the **unused** cells. These give the transportation costs you would expect in these cells, based on the part costs previously calculated.

	11 C1	13 C2	4 C3	5 C4	6 C5	Supplies
−10 P1	36 1	9 3	−6 10	−5 4	−4 7	45
−3 P2	8 — 7	45 10	1 6	2 6	15 3	60
0 P3	11 4	6 13	24 4	45 5	6 7	75
Demands	36	60	24	45	15	180

Cells P2 C1 and P3 C1 are of interest here as the actual transportation costs are lower than the expected ones. Having to spend less than expected is good news, so you should seek to gain advantage from this by using one of these cells. There is a greater possible gain from cell P3 C1 where the cost is £7 per tonne less than expected, so you need to put as large a number as possible here. But how big can it be? Start by introducing an x in P3 C1, then, since the C1 column must total 36, the P1 C1 entry must be reduced to $36 - x$ or C1 will be oversupplied.

Now the P1 row no longer totals 45, and pit P1 has x units of supply to send somewhere. Since sending it to C3, C4 or C5 would mean using cells where the transportation cost is more than the expected cost, these x units should increase P1 C2 to $9 + x$. In moving the x units around the array, apart from the starting point P3 C1, only those cells already employed will be changed. Creating a loop in the array by this idea is sometimes called the **stepping stones method**. Column C2 now has a sum more than 60 so P3 C2 is reduced to $6 - x$ and all the row and column totals are correct. (Reducing P2 C2 instead of P3 C2 would not have been sensible as P2 C5 would then have had to increase to $15 + x$ which it could not as C5's column total prevents it.)

	11 C1	13 C2	4 C3	5 C4	6 C5	Supplies
−10 P1	36−x 1	9+x 3	−6 10	−5 4	−4 7	45
−3 P2	8 7	45 10	1 6	2 6	15 3	60
0 P3	11 x 4	6−x 13	24 4	45 5	6 7	75
Demands	36	60	24	45	15	180

Since all the elements in the solution to the problem must be non-negative (a pit cannot send a negative amount of coal, i.e. **receive** coal from a consumer) the cells P1 C1 and P3 C2 tell you that x can be at most 6. This gives the new array with as much as possible in P3 C1 and none now going from P3 to C2.

	C1	C2	C3	C4	C5	Supplies
P1	30 1	15 3	10	4 −4	7	45
P2	7	45 10	6	6	15 3	60
P3	6 4	13	24 4	45 5	6 7	75
Demands	36	60	24	45	15	180

New cost $= 30 \times 1 + 6 \times 4 + 15 \times 3 + 45 \times 10 + 24 \times 4 + 45 \times 5 + 15 \times 3$
$\qquad = £915$ (thousand). An improvement of £42000!
A fresh set of part costs can now be calculated. An initial assignment of 0 to P3 gives

P1: -3 P2: 4   P3: 0   C1: 4   C2: 6   C3: 4   C4: 5   C5: -1

		C1		C2		C3		C4		C5		Supplies
	4		6		4		5		-1				
-3 P1		$30-x$	1	$15+x$	3	1	10	2		4	-4	7	45
4 P2	8		7	$45-x$	10	8	6	9		6	15	3	60
0 P3	11	$6+x$	4		13		24	4	$45-x$	5	-1	7	75
Demands		36		60		24		45		15		180

Summing the part costs and entering them in the unused cells shows P2 C4 as the greatest source of improvement, so x is inserted here and a loop formed via P3 C4, P3 C1, P1 C1, P1 C2, P2 C2.

Now cells P1 C1, P2 C2 and P3 C4 must not become negative by having too large a value for x, so at most 30 units can be put in P2 C4.

	C1		C2		C3		C4		C5		Supplies
P1		1	45	3	10			4		7	45
P2		7	15	10	6		30	6	15	3	60
P3	36	4		13	24	4	15	5		7	75
Demands	36		60		24		45		15		180

New cost $= 36 \times 4 + 45 \times 3 + 15 \times 10 + 24 \times 4 + 30 \times 6 + 15 \times 5 + 15 \times 3$
$\qquad = £825$ (thousand).

A new set of part costs can now be worked out. Starting with P2: 0 gives

P1: 7   P2: 0   P3: -1 C1: 5   C2: 10   C3: 5   C4: 5   C5: 3

		C1		C2		C3		C4		C5		Supplies
-7 P1	-2		1	45	3	-2	10	-1	4	-4	7	45
0 P2	5		7	15	10	5	6	30	6	15	3	60
-1 P3	36	4	9		13	24	4	15	5	2	7	75
Demands	36		60		24		45		15		

Now no vacant cell invites occupation as each one has an actual transportation cost higher than the expected one. No improvement is possible, so the minimum cost of distribution is £825 000 which is considerably cheaper than the initial solution.

In the final arrangement giving the minimum cost overall, the cheapest cell, P1 C1, is not utilised at all which may, at first, surprising. Using this cell would entail other more expensive routes being employed when supplying C2, C3, C4, and C5, so it is not an economical choice.

Exercise 2.1   Solve the problem posed in Investigation 1.

	S	C	H	Supplies
D1	10	8	9	4
D2	11	3	12	5
Demands	4	3	2	9

Exercise 2.2   Solve the problem in Investigation 2.

	P1	P2	P3	Personnel
B1	34	12	30	20
B2	46	38	41	19
B3	20	44	63	16
Demands	23	14	18	55

(in hundreds of people)

Exercise 2.3   A builder has 25 tonnes of soil at site X, 15 tonnes at site Y and another 15 tonnes at site Z. He has orders for 20 tonnes from customer A, 12 tonnes from customer B, 15 tonnes from customer C and 8 tonnes from customer D. The cost per tonne of moving the soil between the different locations is given in the table below.

		Sites			
		X	Y	Z	
	A	16	6	13	
Customers	B	7	4	10	costs of moving
	C	8	7	9	soil (per tonne)
	D	7	13	9	

How should the builder supply his customers if he is to keep transportation costs to a minimum?

Exercise 2.4   A grocery distributor has three warehouses, W1, W2 and W3, from which to supply four supermarkets, S1, S2, S3 and S4. The supply, demand and unit costs are given in the table. How should the distributor arrange to supply his customers?

	S1	S2	S3	S4	Supplies
W1	9	3	5	3	500
W2	7	13	11	9	600
W3	7	7	5	11	700
Demands	300	400	500	600	1800

9.3   MAXIMISATION PROBLEMS

If the figures in the initial array correspond to profits rather that to costs then the distribution of supplies to meet demands is one to optimise (i.e. maximise) profit. An initial feasible solution can be found by a 'most-profits-first' rule and can be improved by forming expected from part profits where interest is focused on those cells where actual revenues are **higher** than expected.

Such a problem may also be converted to one on minimising by subtracting each profit in the array from the largest. The solution is produced as before and the initial figures used to calculate the final profit.

9.4   UNBALANCED PROBLEMS

It may be that total supply does not equal total demand in some situations. For example, the cost array below gives details of a problem with total demand of 46 units and an overall available supply of 49.

	C1	C2	C3	C4	Supplies
S1	3	2	6	2	11
S2	4	1	4	3	16
S3	7	3	1	2	22
Demands	6	11	17	12	46 / 49

To create a balance between supply and demand a dummy customer, C5, requiring 3 units is introduced. The transportation costs will be the same in each of the cells of C5's column so the supplier associated with the dummy will be determined solely by the figures in the original array. The cost in each cell is zero as no shipment actually takes place, so no expense is involved.

Set up by allocation under the lowest-cost-first rule should use those costs originally given, with any placement in column C5 done last.

Customers

	C1	C2	C3	C4	C5	Supplies
S1	3	2	6	2	0	11
Suppliers S2	4	1	4	3	0	16
S3	7	3	1	2	0	22
Demands	6	11	17	12	3	49

9.5 NON-UNIQUE OPTIMAL SOLUTIONS

The next array shows the stage reached after some work on a transportation problem.

	C1	C2	C3	Supplies
S1	2	5	6 1	6
S2	10 3	4	1 3	11
S3	3	3 3	14 3	17
S4	5	12 1	2	12
Demands	10	15	21	46

Can the solution shown be improved upon?

	0 C1	0 C2	0 C3	Supplies
1 S1	1 2	1 5	6 1	6
3 S2	$10-x$ 3 3	4	$1+x$ 3	11
3 S3	3 x 3	3 3	$14-x$ 3	17
1 S4	1 5	12 1 1	2	12
Demands	10	15	21	46

A comparison of expected and actual costs shows no way in which the answer can be profitably changed but the cell S3 C1 can be utilised without worsening the solution. A loop starting here shows that any value up to 10 can be introduced to the cell, so there are eleven solutions which are optimal, corresponding to $x = 0, 1, ..., 10$.

9.6   DEGENERACY

In a standard transportation problem with m rows and n columns in its array the optimal solution will have $m + n - 1$ entries.

A **degenerate** situation arises when a row and column are exhausted simultaneously in the initial allocation. This results in less than $m + n - 1$ entries and causes problems when calculating part costs and also in forming loops of 'stepping-stones'. The next table's initial solution would be constructed by first putting 3 in cell S3 C1 and then 5 in S2 C2. This second entry both exhausts all S2's supply and satisfies C2's demand. Going on to complete the process would provide only 5 entries where $m + n - 1 = 3 + 4 - 1 = 6$ are required. Before proceeding to put dashes in the row S2 and column C2, enter a zero in one of the cells, e.g. the cheapest, S2 C4. Now continue to form a solution and the number of entries will be 6, as required.

	C1	C2	C3	C4	Supplies
S1	— 14	— 14	3 17	3 16	6
S2	— 13	5 9	— 11	0 10	5
S3	3 8	— 15	2 13	— 13	5
Demands	3	5	5	3	16

Now calculate part costs and expected costs and proceed as usual to look for improvements.

	6 C1	9 C2	11 C3	10 C4	Supplies
6 S1	12 14	15 x 14	3 17	3−x 16	6
0 S2	6 13	5−x 9 11	11	0+x 10	5
2 S3	3 8 11	15	2 13 12	13	5
Demands	3	5	5	3	16

This gives a new non-degenerate solution which can be worked on as usual.

	7 C1	9 C2	12 C3	10 C4	Supplies
5 S1	12 14	3+x 14	3−x 17 15	16	6
0 S2	7 13	2−x 9 12	x 11	3 10	5
1 S3	3 8 10	15	2 13 11	13	5
Demands	3	5	5	3	16

Now $x = 2$ produces an optimal solution which is not unique as can be seen by looking at cell S1 C4.

	-5 C1	-3 C2	0 C3	-1 C4	Supplies
17 S1	12 14	5 14	1 17|16	16	6
11 S2	6 13|8	9	2 11	3 10	5
13 S3	3 8|10	15	2 13|12	13	5
Demands	3	5	5	3	16

Degeneracy can also arise during the course of solving the problem, as illustrated in the following array.

	7 C1	10 C2	7 C3	5 C4	Supplies
-6 S1	15–x 1	45+x 14|1	10|–1	4	60
0 S2	7 7	15–x 10|7	x 6	45 5	60
-3 S3	21+x 4|7	13	24–x 4|2	5	45
Demands	36	60	24	45	165

Now putting $x = 15$ would eliminate both the entries in S1 C1 and S2 C2 and produce a degenerate solution. Putting a zero in one of these cells maintains the $m + n - 1 = 6$ entries required and the solution can proceed. Since S1 C1 is cheaper to utilise it would be natural to keep it in the solution by putting the zero there.

Some optimal solutions are themselves degenerate as can be seen from the next table.

		C1	C2	C3	Supplies
S1		14	13	3 8	3
S2		5 15	10	15	5
S3		17	3 11	2 13	5
S4		16	2 10	13	2
Demands		5	5	5	15

Trying to improve the solution by introducing a zero, for example, to cell S2 C2 and continuing as usual results in the following table, which shows that no alteration can profitably be made.

$_6$S1 $\;$	$_5$ C1	$_0$ C2	$_2$ C3	Supplies
$_6$ S1	11	14 | 6	13 $\quad$ 3 | 8	3
$_{10}$ S2	5 | 15	0 | 10 12	15	5
$_{11}$ S3	16	17	3 | 11 $\quad$ 2 | 13	5
$_{10}$ S4	15	16	2 | 10 12 $\quad$ 13	2
Demands	5	5	5	15

One degelution may lead to another with a different cell being home to the zero before a situation with $m + n - 1 = 6$ entries is reached. If the use of a zero fails to cure the problem and a sequence of degenerate solutions recurs then a different approach called the **perturbation method** is used. (See Section 9.7, *Notes*, for details.)

Exercise 6.1   Milk has to be delivered from farms F1, F2, F3, F4, F5 to dairies D1, D2, D3. The number of tanker loads and distances involved are given in the table. How should the dairies be supplied to minimise the mileage and hence the cost?

		Dairies		
		D1	D2	D3	Supplies
	F1	13	1	19	33
	F2	13	13	16	39
Farms	F3	16	10	13	21
	F4	10	19	22	51
	F5	16	22	19	72
	Demands	57	84	75	216

Comment on the uniqueness of the answer.

Exercise 6.2   A boat hire firm has bookings for six of its fleet on the first day of the holiday season. It has three boats in each of docks D1 and D2 and one boat at dock D3. Customers require one boat at port P1, two at P2 and three at P3. If the unit transportation costs are as given in the table, find the best arrangement for deliveries and its associated cost.

		Ports		
		P1	P2	P3
	D1	17	8	14
Docks	D2	15	10	20
	D3	20	5	10

Exercise 6.3　A steel company has four mills, M1, M2, M3, M4, which can produce 40, 10, 20 and 10, thousand tonnes of steel in a year. Five customers, C1, C2, C3, C4, C5, have requirements for 12, 18, 15, 20 and 15, thousand tonnes respectively in the same period. The costs of production and transportation are shown in the following array. Find the firm's minimum cost in supplying customers.

		C1	C2	Customers C3	C4	C5
	M1	11	9	3	15	3
Mills	M2	7	3	5	9	5
	M3	9	3	3	7	3
	M4	11	5	9	11	9

Costs in £s per tonne

Exercise 6.4　An aggregates firm operates three quarries producing limestone for road-building. These quarries, Q1, Q2, Q3, produce 60, 100 and 80 lorry loads per day. Three sites, S1, S2, S3, need 100, 50 and 60 lorry loads of limestone per day. If the distances are as shown in the table, what is the most economical schedule for distribution?

		S1	Sites S2	S3
	Q1	38	71	23
Quarries	Q2	20	41	5
	Q3	38	83	14

9.7　NOTES

Other terms than those used here may also be encountered; for example,

$$\text{part costs} = \text{shadow costs}$$

$$\text{opportunity costs} = \text{expected costs} - \text{actual costs.}$$

The 'pits and customers' example can be written as a linear programming problem

	C1	C2	C3	C4	C5	Supplies
P1	x_{11}　1	x_{12}　3	x_{13}　10	x_{14}　4	x_{15}　7	45
P2	x_{21}　7	x_{22}　10	x_{23}　6	x_{24}　6	x_{25}　3	60
P3	x_{31}　4	x_{32}　13	x_{33}　4	x_{34}　5 6	x_{35}　7	75
Demands	36	60	24	45	15	180

with objective function $P = x_{11} + 3x_{12} + 10x_{13} + \ldots + 7x_{35}$ to be minimised subject to the seven constraints

(1) $\quad x_{11} + x_{12} + x_{13} + x_{14} + x_{15} \ \leq\ 45$

(2) $\quad x_{21} + x_{22} + x_{23} + x_{24} + x_{25} \ \leq\ 60$ $\quad$ rows

(3) $\quad x_{31} + x_{32} + x_{33} + x_{34} + x_{35} \ \leq\ 75$

(4) $\quad x_{11} + x_{21} + x_{31} \ \leq\ 36$

(5) $\quad x_{12} + x_{22} + x_{32} \ \leq\ 60$ $\quad$ columns

(6) $\quad x_{13} + x_{23} + x_{33} \ \leq\ 24$

(7) $\quad x_{14} + x_{24} + x_{34} \ \leq\ 45$

The equation from C5, $x_{15} + x_{25} + x_{35} = 15$, is not an independent constraint as it can be obtained from the others by taking $(1) + (2) + (3) - (4) - (5) - (6) - (7)$. So, if equations (1) to (7) are satisfied, so is this equation.
Introducing slack variables,

$$x_{11} + x_{12} + x_{13} + x_{14} + x_{15} + y_1 = 45$$
$$x_{21} + x_{22} + x_{23} + x_{24} + x_{25} + y_2 = 60$$
$$x_{31} + x_{32} + x_{33} + x_{34} + x_{35} + y_3 = 75$$
$$x_{11} + x_{21} + x_{31} \qquad\qquad + y_4 = 36$$
$$x_{12} + x_{22} + x_{32} \qquad\qquad + y_5 = 60$$
$$x_{13} + x_{23} + x_{33} \qquad\qquad + y_6 = 24$$
$$x_{14} + x_{24} + x_{34} \qquad\qquad + y_7 = 45$$

and a basic feasible solution is $y_1 = 45$, $y_2 = 60$, $y_3 = 75$, $y_4 = 36$, $y_5 = 60$, $y_6 = 24$, $y_7 = 45$, and $x_{11} = x_{12} = \dots = x_{35} = 0$. Thus you have a solution with $m + n - 1 = 7$ variables.

The variables y_1, y_2, y_3 represent the quantities still to be despatched from pits P1, P2, P3 respectively, and y_4, y_5, y_6, y_7 are the amounts still to arrive at consumers C1, C2, C3, C4. Their initial values correspond to the time before any deliveries are made. Different variables, x_{ij}, may replace them in the solution and some of these may have value zero, but there will continue to be $m + n - 1$.

There are other methods of producing an initial feasible solution. Programmers may prefer a method that starts at a corner (e.g. the top left) and works along each row satisfying customers until supplies are exhausted, as shown in the following array.

	C1	C2	C3	C4	C5	
S1	50	30				80
S2		15	35	10		60
S3				20		20
S4				25	30	55
	50	45	35	55	30	215

Such an allocation is likely to be further from the optimal solution than one produced by the lowest-costs-first rule and so not best for 'pen and paper'.

The **perturbation method** changes the facts of a transportation problem slightly to avoid degeneracy.

	C1	C2	C3	C4	C5	Supplies
S1	60 1	3	5	7	9	60
S2	2	4	6	8	10	70
S3	3	6	9	2	4	80
S4	4	8	5	7	3	90
Demands	60	50	40	75	75	300

The array just given provides a case where degeneracy occurs at once. By adding 0.1 to each supply and 0.4 to C5's demand it becomes impossible to complete a row **and** column until the final entry.

	C1	C2	C3	C4	C5	Supplies
S1	60 1	0.1 3	— 5	— 7	— 9	60.1
S2	— 2	49.9 4	20.2 6	— 8	— 10	70.1
S3	— 3	— 6	5.1 9	75 2	— 4	80.1
S4	— 4	— 8	14.7 5	— 7	75.4 3	90.1
Demands	60	50	40	75	75.4	300.4

At the end of improvements all entries are rounded to the nearest unit – what remains is the optimal solution. With 10 or more suppliers, 0.01 would be needed to replace 0.1.

10

Matching and assignment problems

10.1 INVESTIGATIONS

Investigation 1

Five workers, W1, W2, W3, W4 and W5, are available to do five jobs, J1, J2, J3, J4 and J5. The jobs each worker has the skills to do are shown in the table.

Worker	can do
W1	J1, J2
W2	J2, J5
W3	J1, J3, J4
W4	J1, J4
W5	J2, J4

Is it possible for this team of workers to tackle the five jobs so that everyone is employed?

Investigation 2

Six girls, G1, G2, G3, G4, G5 and G6, from a tennis club want to arrange a mixed doubles tournament with six boys, B1, B2, B3, B4, B5 and B6, whom they know. Fig. 10.1 shows which boys each of the girls is prepared to partner.

Can they be arranged into six pairs for the competition?

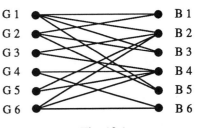

Fig. 10.1

10.2 HALL'S MARRIAGE THEOREM

The **'marriage problem'**, as it is known, poses the question, "Is it possible to marry a finite set of boys to a finite set of girls so that each one is paired with one he knows"?

For example, if there are five boys who know six girls as detailed in the table, can the boys be (monogamously) married to girls they know?

Boy	knows
B1	G1, G2, G4, G6
B2	G2, G4, G5
B3	G1
B4	G1, G3, G5
B5	G2, G3, G6

The information may also be displayed as a **bipartite network** (Fig. 10.2), that is one in which the nodes can be divided into two sets so that none of them is connected to another in the same set.

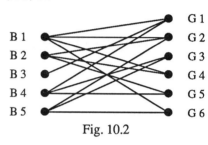

Fig. 10.2

The problem can be solved by the matching B1, G4; B2, G5; B3, G1; B4, G3; B5, G2; other solutions are also possible.

Hall's theorem states that the problem can be solved if and only if every set of n boys, for $n = 1, 2, 3, 4, 5,$ knows between them at least n girls. In other words, every group of two boys must know at least two girls; each group of three boys needs to know three or more girls, etc.

Investigation 2 had no solution as the group of four girls, G3, G4, G5 and G6, were only prepared to partner a group of three boys, B2, B4 and B6.

10.3 MATCHING IMPROVEMENT ALGORITHM

Rather than trying to test the feasibility of solving a problem by applying Hall's theorem to the many possible groups of 2, 3, etc., it is usually more practical to find a partial matching and then try to improve it.

Example

Is it possible to match each of A, B, C, D and E with a different node from the set 1, 2, 3, 4 and 5 using only the arcs shown in Fig. 10.3?

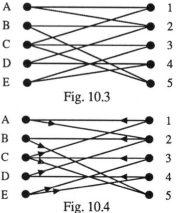

Fig. 10.3

An initial partial matching could be A1, B2, C3, D4 which leaves E and 5 unpaired. This partial matching is indicated by arrows from right to left while the free arcs are given directions from left to right. (Fig. 10.4)

Fig. 10.4

It is convenient to put the arrows near the tail
nodes.

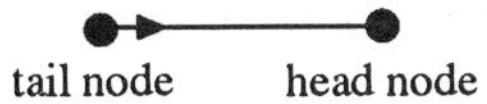

tail node   head node

The problem now is to adjust this initial
arrangement so that E may also be matched.
If E is to be matched, which are the candidate
nodes in the right-hand set? The left-to-right
arrows tell you that 3 and 4 could be connected
to E. (Fig. 10.5)

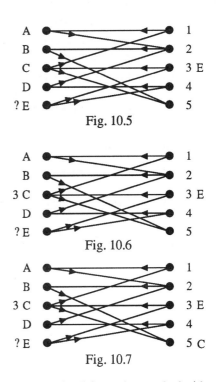

Fig. 10.5

Now if 3 is to be freed for matching with E it
must be disconnected from its present pairing.
Which is it paired with? The right-to-left
arrow shows that 3 is presently matched with
C. (Fig. 10.6)

Fig. 10.6

Which other nodes on the right could C be
paired with? The left-to-right arrows tell you
that C could be paired with 5. (Fig. 10.7)

Fig. 10.7

A possible re-pairing has emerged. C may be linked with 5, leaving 3 free to be matched with
E. This gives a complete matching

A1, B2, C5, D4, E3.

The pairing 3C shown on the left of the diagram (Fig. 10.7) gives the combination to be
discarded from the original partial matching and the ones on the right are those to come in.

The method then is based on looking for a route from E to 5 passing along the arcs only
in the directions indicated. Such a route cannot pass through the same node twice. If, for
example, it went to node 3 twice, this would mean that you wanted to pair two of the nodes
on the left with it. In this case, the route was E3, C5 and writing the nodes in pairs, E3, C5
gives the new pairing to improve the initial partial matching. Failure to find a route from E
to 5 via 3 would mean that one through 4 would be sought. If it had not been possible to find
such a route from E to 5 then a complete matching would not have existed.

This may be formalised as an algorithm.

Step 1   Show the information as a bipartite network.

Step 2   Find the partial matching. Indicate all arcs involved in this partial matching
with arrows from right to left. Put arrows from left to right on all the free arcs.

Step 3   Label an unmatched node in the left set with '?'.

Step 4   Label any unlabelled head node in the right set which is connected to the most recently labelled tail nodes in the left set by noting the tail node by it. If no such node exists go to Step 7.

Step 5   Label any unlabelled head node in the left set which is connected to the most recently labelled tail nodes in the right set by noting the tail node by it. If no such node exists go to Step 7.

Step 6   Repeat Steps 5 and 6 until a right hand node is reached which was not in the initial partial matching. Note the nodes on the path and write them in order in pairs. Add to these any original pairings whose right to left arcs are not part of the path.

Step 7   Repeat Steps 3 to 6 if there are any further unmatched nodes not yet considered.

Step 8   STOP when all nodes are matched or matchings have been sought.

Note: If there is more than one label that may be given to an unlabelled node then any of those available may be used.

The path found by this method is often called an **alternating path** as the nodes on it come alternately from the two sets, left and right.

Exercise 3.1   Use the matching improvement algorithm and the given initial partial matching to try to find as many complete matchings as possible in each of the following groups.

(a)   Partial matching

A1, B3, C5, D2

(b)   Partial matching

A2, B1, C3, D4

(c)   Partial matching

A1, B2, E5, F6

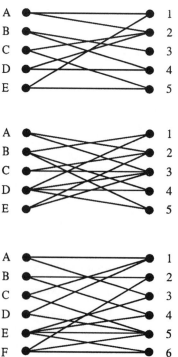

(d)   Partial matching

     A1, C3, D4, E5

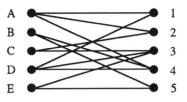

Exercise 3.2   At a bridge club a mixed pairs competition is being arranged to involve Mr. A, Mr. B, Mr. C and Mr. D and Ms. W, Ms. X, Ms. Y and Ms. Z.

Mr. A will not partner Ms. W; Mr. B will only be in a team with Ms. W or Ms. Y; Mr. C will partner Ms. W or Ms. Z and Mr. D is prepared to be in a team with anyone except Ms. W.

If the ladies are less awkward and do not mind whom they partner, find as many arrangements for the tournament as possible.

Exercise 3.3   A college has to staff classes in Russian, German, Spanish, Italian, Dutch and French and has six language teachers available.

Mr. A can teach Russian, German or Dutch; Mrs. B is only qualified to take classes in Spanish or Italian; Mr. C can teach only French or Spanish; Mrs. D does not teach Spanish or Italian; Mr. E can take lessons in Spanish, Italian or French; Ms. F teaches only French and Italian.

How can the six courses be staffed?

Exercise 3.4   Four runners need to be organised into a 4×100 m relay team. Ann can run 1st, 2nd or 4th legs. Beth runs 2nd or 3rd. Clare goes 1st or 2nd. Denise likes to run either of the last two legs. In how many ways can the team be arranged?

If their best times for the different legs are as detailed in the table, which order is best?

	leg			
	1	2	3	4
Ann	12.3	11.9	12.1	11.8
Beth	12.1	11.7	11.8	11.7
Clare	12.2	12.0	12.0	12.1
Denise	12.3	11.8	12.0	11.9

Exercise 3.5   Four experienced sailors, Mrs. Salt, Mr. Cruise, Miss Brine and Mr. Tarr, are to take out eight novices for lessons, with two novices going with each experienced sailor. As a result of the different times when people are available, only some arrangements are possible. The possibilities are shown in the table.

	novices							
	1	2	3	4	5	6	7	8
Mrs. Salt	✔	✔		✔			✔	
Mr. Cruise		✔	✔		✔			✔
Miss Brine	✔			✔		✔		
Mr. Tarr			✔		✔		✔	✔

Starting with the partial matching S14, C23, B6, T78, use the matching improvement algorithm to find a feasible solution.

Hint: Let each experienced sailor be represented by two nodes. To make the network simpler the second set representing them could be drawn on the other side of the set representing the novices. When more than one matching is possible there may be a 'best' pairing.

10.4 INVESTIGATIONS
Investigation 3

Three cyclists, Leif, Mark and Nathan, are to ride as a team in an event which requires one of them to do a 10 km road race, another to do a 5 km track race and the third to compete in a cross-country section. If the winning team is the one with the smallest total time, how should they decide who does which ride to give themselves the best chance? Their practice times are given below in minutes to the nearest minute.

	10 km	track	cross-country
Leif	20	8	15
Mark	19	6	12
Nathan	22	9	17

Investigation 4

A family wants to have several jobs done on their new house in the two days immediately before they move in. They get estimates from four small firms who will each be given one part of the work so that it can be completed in time. The prices given by the firms are shown in the table. What is the cheapest way to distribute the jobs?

	jobs			
	A	B	C	D
Firm 1	100	190	180	110
Firm 2	130	160	120	140
Firm 3	180	220	170	150
Firm 4	110	240	190	100

costs in £

Job A :   Fit new wardrobes and redecorate main bedroom.

Job B :   Remove old units, instal new ones and redecorate the kitchen.

Job C :   Repaint remaining six rooms.

Job D :   Repaint the exterior of the house.

10.5 HUNGARIAN ALGORITHM
Example

Suppose that five people, P1, P2, P3, P4 and P5, can do each of the five jobs, J1, J2, J3, J4 and J5, in the times, given in minutes, in the table below.

	J1	J2	J3	J4	J5
P1	13	8	12	21	14
P2	17	23	10	16	18
P3	14	13	15	15	16
P4	17	8	11	16	14
P5	12	7	15	20	11

Step 1   A constant may be added to or subtracted from every element in a row without altering the optimal arrangement. For example, subtracting 8 from each entry in the first row simply means that you are now noting how much over eight minutes P1 takes on each task instead of the total time and this will not effect which job the person is finally assigned to.

Subtract 8 from the first row, 10 from the second, 13 from the third, 8 from the fourth and 7 from the fifth. The table becomes

	J1	J2	J3	J4	J5
P1	5	0	4	13	6
P2	7	13	0	6	8
P3	1	0	2	2	3
P4	9	0	3	8	6
P5	5	0	8	13	4

Now every row has a zero, which indicates each person's best job but as these zeros are in just two columns, you cannot assign everyone to the job they would choose.

Step 2   This process of reducing by subtraction can also be applied to columns. Again optimality is not affected as you are simply measuring how much more than a set time a particular job takes each person. From each entry in a column, take the smallest number in it. In this case take 1 from the first column, 0 from the second and third columns, 2 from the fourth and 3 from the fifth column to give the table as shown.

	J1	J2	J3	J4	J5
P1	4	0	4	11	3
P2	6	13	0	4	5
P3	0	0	2	0	0
P4	8	0	3	6	3
P5	4	0	8	11	1

These two reductions by rows and columns can be performed in the reverse order. The resulting figures are different if this is done but the final solution is the same. It is quite possible that using one order rather than the other will reduce the amount of work needed to solve a particular problem completely but neither one is always quicker than the other.

Step 3   There is still no immediately obvious way to assign the jobs to the five people. Looking at the first row, the zero indicates that it would be a good idea to assign J2 to P1 but if row 1 and column 2 are taken out of consideration you have no zeros in the bottom two rows to suggest how P4 and P5 should be matched with jobs. This happens as so many of the zeros appear in the same column. Only when the minimum number of horizontal and/or vertical lines needed to cover the zeros is five will a complete assignment be found. In general then, for n people and tasks you require n lines.

The smallest number of lines needed to cover all the zeros in this case is three, and since this is less than five you have reached a solution to the problem.

	J1	J2	J3	J4	J5	
·P1	4	0	4	11	3	
→ P2	6	13	0	4	5	←
→ P3	0	0	2	0	0	←
P4	8	0	3	6	3	
P5	4	0	8	11	1	

Step 4   Find the smallest entry not covered by the lines. In this example P5, J5 = 1 is the least. Subtract 1 from each uncovered element and add 1 to any that is covered twice. This is equivalent to adding 1 to each covered row and subtracting 1 from each uncovered column. Neither of these operations changes optimality, as discussed in Steps 1 and 2. This produces the new table shown below, for which a minimum of four lines are now needed to cover the zeros. Five are still not required so a complete assignment is still not found, though it is closer.

	J1	J2	J3	J4	J5	
·P1	3	0	3	10	2	
P2	6	14	0	4	5	
→ P3	0	1	2	0	0	←
P4	7	0	2	5	2	
P5	3	0	7	10	0	

Now repeat Step 4. The minimum value not covered by the four lines is P1, J1 = 3, so subtract 3 from each uncovered element and add 3 to any element covered twice.

	J1	J2	J3	J4	J5
P1	0	0	3	7	2
P2	3	14	0	1	5
P3	0	4	5	0	3
P4	4	0	2	2	2
P5	0	0	7	7	0

This new table requires five lines to cover all its zeros so an allocation of jobs is now possible.

Step 5   Look for any rows or columns containing a single zero as they indicate forced assignments. Since P2 contains just one zero the allocation of P2 to job J3 is fixed. Cover the row P2 and column J3 and repeat the process of seeking forced matches. Here this leads to the complete assignment

P1, J1   P2, J3   P3, J4   P4, J2   P5, J5

which, according to the original data, gives a minimum total of

$$13 + 10 + 15 + 8 + 11 = 57 \text{ minutes.}$$

Just as rows and columns can have numbers added to or subtracted from them, they may also be multiplied or divided by numbers. For example, multiplying by ten could be used to eliminate decimals. These changes are equivalent to a change of units for the numbers in the table.

No method has been included here to decide where to put the lines to cover the zeros. This is the part of the algorithm that has not been spelt out as a specific method and is only needed for very large arrays or writing a computer program. One is given section 10.10 *Notes*, at the end of this chapter.

Exercise 5.1   Re-do the example just covered, reducing columns before rows.

	J1	J2	J3	J4	J5	
P1	13	8	12	21	14	
P2	17	23	10	16	18	
P3	14	13	15	15	16	*times in minutes*
P4	17	8	11	16	14	
P5	12	7	15	20	11	

Exercise 5.2   Use the Hungarian algorithm to solve the problem set in Investigation 4.

	A	B	C	D	
Firm 1	100	190	180	110	
Firm 2	130	160	120	140	
Firm 3	180	220	170	150	*costs in £*
Firm 4	110	240	190	100	

Exercise 5.3   A company has depots at Mancastle, Newbridge, Camdon and Lonchester
and a single lorry available at each depot. New supplies are required at its shops in Oxfield,
Sheffingham, Nottol and Brisford and any depot has sufficient in stock to supply any shop.
If the distances are as given in the table, which depot should supply which shop?

	Oxfield	Sheffingham	Nottol	Brisford	
Mancastle	100	97	146	83	
Newbridge	47	51	72	40	*distances in*
Camdon	126	130	153	104	*miles*
Lonchester	78	90	83	69	

Exercise 5.4   Each summer a firm employs five students to help cover for holiday
absences. This year Ashoke, Fiona, Gemma, Karl and Tamsin have been given jobs and need
to be trained before starting work. Bearing in mind the previous experiences of the students
the manager assesses the likely training costs to be as given in the table.

	driver	packer	accounts clerk	computer operator	secretary/ receptionist	
Ashoke	90	85	90	120	70	
Fiona	90	130	125	120	100	
Gemma	80	115	120	100	115	*costs*
Karl	140	160	140	135	125	*in £*
Tamsin	95	125	150	130	110	

How should the training be organised to minimise the overall cost?

10.6 IMPOSSIBLE ASSIGNMENTS
Example

A medley relay team consists of Helen, Jenny, Ruth and Sireesha. They are required to swim
crawl, backstroke, breaststroke and butterfly. Each of them can swim any of the strokes,
except Helen, who does not do the butterfly. Their times for the different strokes are given
in the table.

	crawl	breaststroke	backstroke	butterfly
Helen	37	42	44	*
Jenny	39	42	46	50
Ruth	38	40	40	49
Sireesha	41	41	42	51

times in seconds

Which leg of the relay should each girl swim if their total time is to be as small as possible?

Solution

Throughout, the solution * is taken as being infinitely large. Subtract 37 from the first row, 39 from the second, 38 from the third and 41 from the fourth.

	crawl	breaststroke	backstroke	butterfly
Helen	0	5	7	*
Jenny	0	3	7	11
Ruth	0	2	2	11
Sireesha	0	0	1	10

Now take 1 from the third column and 10 from the fourth.

	crawl	breaststroke	backstroke	butterfly
Helen	0	5	6	*
Jenny	0	3	6	1
Ruth	0	2	1	1
→ Sireesha	0	0	0	0 ←

The minimum number of lines $= 2 < 4$ and the least uncovered element is 1, so take 1 from each uncovered cell and add 1 to the one covered twice.

	crawl	breaststroke	backstroke	butterfly
Helen	0	4	5	*
Jenny	0	2	5	0
Ruth	0	1	0	0
Sireesha	1	0	0	0

Now four lines are needed to cover the zeros so swimmers can be assigned to the different strokes.

Helen must do the crawl and the breastroke must be done by Sireesha. Covering these two rows and columns you can see that Jenny will swim the butterfly and Ruth will take the backstroke leg, giving a total time of

$$37 + 41 + 50 + 40 = 168 \text{ seconds.}$$

10.7 MAXIMISING PROBLEMS

If the values in the table represent profits, then the problem becomes how to match people to jobs so as to maximise profits. Consider the problem represented by the following table.

	J1	J2	J3	J4	J5
P1	7	13	6	11	6
P2	9	10	9	8	4
P3	9	10	9	7	6
P4	6	3	8	4	3
P5	9	6	12	8	5

The solution now depends on selecting large numbers, whereas before the method was based on the creation of small numbers, in particular, zero. Multiplying all the elements in the array by minus one makes the largest number, 13, into the smallest, –13. Now adding 13 to each one removes all the negatives and a matrix has been created from the original so that the largest numbers in the original have been transformed into the smallest in the new version.

This can be summarised in a simple operation. Take each element in the original array from the largest one.

	J1	J2	J3	J4	J5
P1	6	0	7	2	7
P2	4	3	4	5	9
P3	4	3	4	6	7
P4	7	10	5	9	10
P5	4	7	1	5	8

Now using the Hungarian algorithm will identify the cells to use and the original table is used to calculate the maximum profit.

If a maximising problem contains an impossible assignment the * remains unchanged throughout these operations.

Exercise 7.1   Maximise

	J1	J2	J3
P1	11	1	5
P2	3	7	14
P3	9	9	14

Exercise 7.2   Minimise

	J1	J2	J3	J4
P1	6	8	11	9
P2	12	*	13	13
P3	10	10	17	14
P4	13	11	13	11

Exercise 7.3   Ann, Carol, Denise, Thani and Wendy will each do one event from high jump, long jump, 100 metres, hurdles and 200 m in the next athletics match. Carol does not hurdle. The number of points they have averaged in these events so far this season are given in the table.

	high jump	long jump	100 m	hurdles	200 m
Ann	4.1	3.9	2.7	4.4	1.9
Carol	3.2	3.4	3.9	*	3.2
Denise	1.9	3.3	3.6	2.5	3.1
Thani	2.3	3.4	3.7	3.6	3.0
Wendy	3.8	3.8	2.8	4.0	2.4

Who would you recommend for each event?

10.8 NON-UNIQUE SOLUTIONS

Suppose the supply and demand question, Exercise 5.3, has a slight change made to one of its entries. A new road has been opened which reduces the distance from Oxfield to Mancastle by seven miles. How does this affect the problem's solution?

	Oxfield	Sheffingham	Nottol	Brisford
Mancastle	93	97	146	83
Newbridge	47	51	72	40
Camdon	126	130	153	104
Lonchester	78	90	83	69

reduce by rows:				
	10	14	63	0
	7	11	32	0
	22	26	49	0
	9	21	14	0

↓

reduce by columns:	3	3	49	0
→	0	0	18	0←
needs 3 lines to cover the zeros	15	15	35	0
→	2	10	0	0←

↑

3 is the smallest uncovered element, so reduce those not covered by 3 and increase those covered twice by the same amount.

	Oxfield	Sheffingham	↓ Nottol	↓ Brisford
Mancastle	0	0	46	0
Newbridge	0	0	18	3
→ Camdon	12	12	32	0 ←
→ Lonchester	2	10	0	3 ←
			↑	↑

Now four lines are required to cover the zeros so an assignment pattern can be found. Camdon – Brisford and Lonchester – Nottol are forced matches but deleting their rows and columns you see that two assignments are possible for those remaining.

The two assignment patterns that minimise the total distance are

(i) Camdon – Brisford
    Lonchester – Nottol
    Mancastle – Oxfield
    Newbridge – Sheffingham

(ii) Camdon – Brisford
    Lonchester – Notol
    Mancastle – Sheffingham
    Newbridge – Oxfield

10.9 UNBALANCED PROBLEMS
Example

The school athletics team in Exercise 7.3 has a new member who has just moved to the area. She is a better hurdler than any of the others so the five girls are now in competition for just four places in the team. Which event should each girl do and who will be left out?

	high jump	long jump	100 m	200 m
Ann	4.1	3.9	2.7	1.9
Carol	3.2	3.4	3.9	3.2
Denise	1.9	3.3	3.6	3.1
Thani	2.3	3.4	3.7	3.0
Wendy	3.8	3.8	2.8	2.4

Solution

This is a maximising problem so multiply all the figures by ten to remove the decimals and subtract them from the largest, which will be 41.

	high jump	long jump	100 m	200 m
Ann	0	2	14	22
Carol	9	7	2	9
Denise	22	8	5	10
Thani	18	7	4	11
Wendy	3	3	13	17

Balance the number of people and events by introducing a 'dummy' event and ensure no one athlete is more likely than any other to be assigned to it by making all the entries in its column the same. As you can reduce columns first you might as well choose zeros for the last column as any other choice will give the same result after one step.

	high jump	long jump	100 m	200 m	–
Ann	0	2	14	22	0
Carol	9	7	2	9	0
Denise	22	8	5	10	0
Thani	18	7	4	11	0
Wendy	3	3	13	17	0

Reduce by columns:

	high jump	long jump	100 m	200 m	–
→ Ann	0	0	12	13	0 ←
→ Carol	9	5	0	0	0 ←
Denise	22	6	3	1	0
Thani	18	5	2	2	0
Wendy	3	1	11	8	0

The smallest uncovered element is one, so take this from each not covered and add one to those covered twice.

	high jump	long jump	100 m	200 m	–
Ann	0*	0	12	13	1
Carol	9	5	0*	0	1
Denise	21	5	2	0	0
Thani	17	4	1	1	0*
Wendy	2	0	10	7	0

Now five lines are needed and three assignments are forced as indicated by *. When the three rows and columns containing * are deleted, the last two pairings can be made, giving

Ann – high jump   Carol – 100 m
Denise – 200 m   Wendy – long jump

and Thani, having been paired with the dummy event, has no place in the team.

Exercise 9.1 Five maths teachers, Mr. Long, Miss Evans, Mrs. Newton, Ms. Philips and Mr. Francis, are available to teach new sixth form sets next year. Four sets are available. One set needs a pure maths teacher while the second will study mechanics, the third statistics and the fourth will follow a discrete maths course. If the numbers in the table give the percentage of the teachers' candidates who have failed the different courses in recent years, how should the head of maths assign the teachers to minimise the likely number of fail grades?

	Pure	Mechanics	Statistics	Discrete
Mr. Long	5.0	4.2	4.8	7.3
Miss Evans	2.3	2.0	2.5	3.6
Mrs. Newton	3.9	3.4	4.5	4.1
Ms. Philips	6.3	5.2	6.5	7.6
Mr. Francis	4.3	4.6	3.0	6.2

Exercise 9.2 A football club manager has put five of his players up for sale and has received a number of bids as shown in the table. The players are named Pinder, Cooper, Jones, Patel and Adams.

	Pinder	Cooper	Jones	Patel	Adams
Aston City	25	36	35	41	41
Sheffield Friday	43	25	42	31	49
Chelocean	34	38	37	43	*
East Bromwich	21	37	36	40	30
West Spam	40	30	41	32	45

All figures are in thousands of pounds . * indicates that no offer has been made. Which players should be sold to which clubs if the manager is to bring in the maximum amount of money possible?

Exercise 9.3 Five expert machine operators can handle any of the five machines available and a job is only complete when it has passed through all five. When Rachel left the firm, the others, Naomi, Deborah, Sarah and Ruth, could work any four machines, with a new operator being trained for the fifth. If the operating times for the machines are as given in the table, which of the machines one to five should the newcomer be trained for?

	machine 1	machine 2	machine 3	machine 4	machine 5
Naomi	51	42	29	33	48
Deborah	44	39	38	39	43
Sarah	33	45	42	43	34
Ruth	34	47	46	45	39

times in minutes

Exercise 9.4   A wargamer has four attacking formations, A1, A2, A3, and A4, available to assault his enemy's four defensive positions, D1, D2, D3, D4. He estimates the various possible casualties his forces could suffer as shown in the matrix. How should the attack be organised to minimise casualties?

	D1	D2	D3	D4
A1	170	100	160	*
A2	130	*	150	160
A3	80	110	160	140
A4	70	150	*	110

10.10 NOTES
It is possible to set out assignment problems in the same sort of table as used for transportation. Since each person is to be assigned to just one task, the 'supply' and 'demand' are one in every row and column. Thus the problem comes under the heading of **degenerate** as there are n entries rather than $2n - 1$ with n jobs and n people.

LINE SELECTION
Example

How many lines are needed to cover the zeros in the table given?

	J1	J2	J3	J4	J5
P1	3	0	12	1	8
P2	9	2	2	3	0
P3	0	1	0	0	2
P4	11	0	3	5	5
P5	3	0	7	0	3

Solution

Step 1   Make as many assignments as possible. Four can be made here, P1J2, P2J5, P3J1 and P5J4.

(Other selections of four are possible.)

Mark these zeros $\bar{0}$.

	J1	J2	J3	J4	J5
P1	3	$\bar{0}$	12	1	8
P2	9	2	2	3	$\bar{0}$
P3	$\bar{0}$	1	0	0	2
P4	11	0	3	5	5
P5	3	0	7	$\bar{0}$	3

	J1	J2	J3	J4	J5
P1	3	$\bar{0}$	12	1	8
P2	9	2	2	3	$\bar{0}$
P3	$\bar{0}$	1	0	0	2
P4	11	0	3	5	5 ✔
P5	3	0 ✔	7	$\bar{0}$	3

Step 2 Put a tick (✔) by each row which has not been assigned.

Step 3 Put a tick by each column containing a zero in a ticked row.

	J1	J2	J3	J4	J5
P1	3	$\bar{0}$	12	1	8 ✔
P2	9	2	2	3	$\bar{0}$
P3	$\bar{0}$	1	0	0	2
P4	11	0	3	5	5 ✔
P5	3	0 ✔	7	$\bar{0}$	3

Step 4 Put a tick by a row which has a $\bar{0}$ in a column which is ticked.

Step 5 Repeat Steps 3 and 4 until no more ticks can be added.

Step 6 The rows without ticks and the columns with ticks provide a minimum set of lines to cover the zeros; in this case four.

Even this method lacks precision as it requires the correct number of assignments to be made in Step 1. Computer programmers may be well advised to try an altogether different approach, such as Mack's Method.

LINEAR PROGRAMMING

An assignment problem may be formulated in such a way as to be solvable using **linear programming**.

	J1	J2	J3
P1	8	6	5
P2	3	2	9
P3	7	1	4

Example

The table shown may be put in the same form used for transportation problems.

This gives rise to five constraints.

	J1	J2	J3	
P1	x_{11} 8	x_{12} 6	x_{13} 5	1
P2	x_{21} 3	x_{22} 2	x_{23} 9	1
P3	x_{31} 7	x_{32} 1	x_{33} 4	1
	1	1	1	3

rows :
$$x_{11} + x_{12} + x_{13} = 1 \qquad (1)$$
$$x_{21} + x_{22} + x_{23} = 1 \qquad (2)$$
$$x_{31} + x_{32} + x_{33} = 1 \qquad (3)$$
columns :
$$x_{11} + x_{21} + x_{31} = 1 \qquad (4)$$
$$x_{12} + x_{22} + x_{32} = 1 \qquad (5)$$

Objective function:

$$P = 8x_{11} + 6x_{12} + 5x_{13} + 3x_{21} + 2x_{22} + 9x_{23} + 7x_{31} + x_{32} + 4x_{33}$$

to be minimised/maximised.

The column constraint $x_{13} + x_{23} + x_{33} = 1$ can be included but is not independent and so contains no new information. It is formed by $(1) + (2) + (3) - (4) - (5)$ so if they are satisfied it will be, automatically.

11

Game theory

11.1 INVESTIGATION

You are playing a game against a machine and have three strategies, p1, p2, p3, you can adopt each time you play. The machine has two plays it can use, called m1 and m2. The outcomes when different combinations of tactics are employed are shown in the table. So if you play p2 and the machine uses m1 you lose four points, but were the machine to have adopted m2 instead you would have gained three points.

Would you use the same plan each time or adopt a mixture of strategies? If you employ a combination, which ones are involved and on what fractions of the plays would you use them?

		Machine	
		m1	m2
Player	p1	6	–2
	p2	–4	3
	p3	–2	–1

Swali

11.2 THE MINIMAX THEORY

Two players, A and B, play the following game which results in this pay-off matrix. Player B has three possible strategies, b1, b2 and b3, while player A could adopt plan a1 or plan a2.

		Player B		
		b1	b2	b3
Player A	a1	(3, –3)	(–5, 5)	(2, –2)
	a2	(–2, 2)	(3, –3)	(–1, 1)

In the array (4, –4) means that A wins 4 and B wins –4, that is, A wins 4 and B loses 4.

The entries in each element of the matrix add to zero since what one player gains the other loses. This is an example of a **zero-sum game**.

Since the second entry in each cell is determined by the first, you need only note one, for example the first, so the matrix becomes

Player B

		b1	b2	b3
	a1	3	–5	2
Player A	a2	–2	3	–1

–3   –3   –2

Each figure represents Player A's winnings and the negative of each gives Player B's winnings.

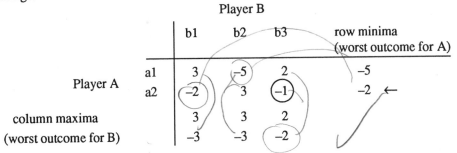

Player B

		b1	b2	b3	row minima (worst outcome for A)
	a1	3	–5	2	–5
Player A	a2	–2	3	–1	–2 ←
column maxima (worst outcome for B)		3	3	2	
		–3	–3	–2	

If A plays safe and plans a strategy on the assumption that the worst scenario will occur, then plan a2 will be the choice as it has the smallest maximum possible loss (MINIMAX strategy). Similarly B should go for b3. The result of A playing a2 while B tries b3 is that A will lose one point and B will gain one.

If A knows or assumes that B will always play safe (i.e. adopt b3), will a switch from a2 to a1 be advantageous? Yes, since a1, b3 gives +2 instead of –1.

What about B's best plan? If B assumes that A will adopt strategy a2 is it a good idea to switch from b3? Yes, B would do well to switch to plan b1 where the gain will increase from 1 to 2. If both players change their strategies the result will be 3 for A (cell a1 b1) instead of 1 for B.

The solution provided by the minimax strategy was cell a2 b3 but in this case it is said to be unstable, so there is an incentive for (one or) both players to change from the play-safe strategy. (This is a **minimax mixed** strategy.)

In the next array the cell a2 b2 has been selected.

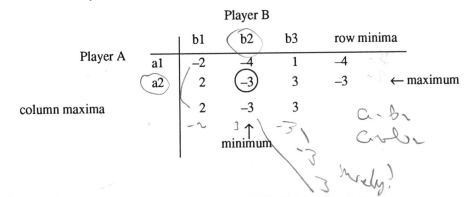

Player B

		b1	b2	b3	row minima
Player A	a1	–2	4	1	–4
	a2	2	–3	3	–3 ← maximum
column maxima		2	–3	3	
			3 ↑ minimum		

You can see that it does not pay B to switch to either b1 or b3 as losses are greater in a2 b1 and a2 b3. Nor does it help A to change to a1 as a1 b2 is worse than a2 b2. The situation here has a stable solution, known as a **saddle point**.

Exercise 2.1   Decide whether or not the following zero sum games have stable solutions.

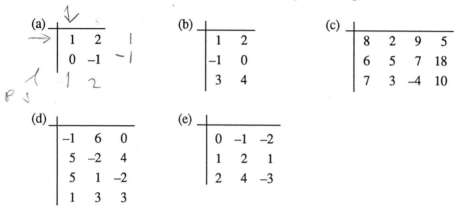

(a)

1	2
0	−1

(b)

1	2
−1	0
3	4

(c)

8	2	9	5
6	5	7	18
7	3	−4	10

(d)

−1	6	0
5	−2	4
5	1	−2
1	3	3

(e)

0	−1	−2
1	2	1
2	4	−3

Exercise 2.2 Compare each stable solution with other numbers
(a)   in the same row,
(b)   in the same column.

What do you notice?

11.3 WHAT IS A ZERO-SUM GAME?
A **zero-sum game** has a scoring system such that whatever one player wins, the other loses.

This matrix is a description of a zero-sum game, as it satisfies the criterion above.

		Player B		
		b1	b2	b3
Player A	a1	(4, 0)	(3, 1)	(1, 3)
	a2	(2, 2)	(1, 3)	(0, 4)
	a3	(1, 3)	(0, 4)	(2, 2)

This becomes clearer if we subtract two points from each player's score. Now each cell in the array has a sum of zero, but are the two systems equivalent?

		Player B		
		b1	b2	b3
Player A	a1	(2,−2)	(1, −1)	(−1, 1)
	a2	(0, 0)	(−1, 1)	(−2, 2)
	a3	(−1, 1)	(−2, 2)	(0, 0)

Under the first scheme three games could result in
(4, 0) (3, 1) (1, 3)   ⟹   A leads by 4, (8 − 4)

The second method of scoring gives
$$(2, -2)\ (1, 1)\ (-1, 1)\ \Rightarrow\ \text{A leads by } 4\ (2 - (-2))$$
There are many examples of zero sum games out in the world. Every time you serve to play a point in table tennis you start a zero sum game, since either you will win the point and your opponent will lose it $(1 - 0)$ or vice versa $(0 - 1)$.

This is not true in English league soccer where points are distributed, $3 - 0$, $1 - 1$ or $0 - 3$, but is true in some other leagues.

11.4 STABLE SOLUTIONS

Consider the following pay-off matrix for players A and B where the entries represent the amounts that B pays to A.

		Player B strategies				
		b1	b2	b3	b4	row minima
Player A strategies	a1	8	2	9	15	2
	a2	6	5	7	18	5 ←max (α)
	a3	7	3	–4	10	–4
column maxima		8	5	9	18	

$$\uparrow$$
$$\min\ (\beta)$$

The element shown in the square is the smallest element in its row and the largest in its column so there is no profit for either A or B in changing strategy. This is an example of a **minimax pure strategy**.

The figure in the box is the **value** of the game, that is, the pay-off, for player A.

For any zero-sum game $\alpha \le \beta$.

A zero-sum game has a stable solution if and only if $\alpha = \beta$

α is the maximin, i.e. the maximum of the row minima.

β is the minimax, i.e. the minimum of the column maxima.

Can a zero-sum game have more than one stable solution (saddle-point)?

The previous matrix with the first column changed gives an array with two saddle-points, each of which gives a value of 5.

		Player B				
		b1	b2	b3	b4	row minima
Player A	a1	4	2	9	15	2
	a2	5	5	7	18	5 ← max
	a3	–1	3	–4	10	–4
column maxima:		5	5	9	18	

$$\uparrow\qquad\uparrow$$
$$\min\qquad\min$$

The **trivial matrix** has six stable solutions

		b1	b2	b3
			Player B	
Player A	a1	1	1	1
	a2	1	1	1

11.5 ROW AND COLUMN DOMINATION
Consider the zero-sum game with pay-off matrix given here.

		b1	b2	b3	b4
			Player B		
Player A	a1	8	7	4	-3
	a2	9	7	6	1
	a3	-12	10	-5	-9

In this case, row two is said to dominate row one since it gives a better pay-off for the first player, regardless of the strategy adopted by the second. The array can, therefore, be reduced as shown since strategy a1 will never be adopted.

		b1	b2	b3	b4
			Player B		
Player A	a2	9	7	6	1
	a3	-12	10	-5	-9

Now look at the columns giving the pay-offs for the second player (negatively speaking) who is looking for low values. The fourth column has values which are lower than those in both b2 and b3, so these two will not be played and the pay-off matrix then reduces to a two by two array.

		b1	b4
		Player B	
Player A	a2	9	1
	a3	-12	-9

But now you can reject a3 to give

		b1	b4
		Player B	
Player A	a2	9	1

and then drop b1 so all that remains is

		b4
Player A	a2	1

A matrix will only reduce this far when the game has a stable solution, but it is always worth looking at row and column domination.

Since zero-sum games can be written as linear programming problems, a simplification could mean that the problem may be solved by graphical methods without the need for a simplex tableau.

Exercise 5.1   Determine whether or not the following matrices can be reduced, and make whatever simplifications are possible.

(a)

3	5	4
1	4	2
6	3	7

(b)

2	1	5	-3
4	2	2	1
3	-1	1	2

(c)

2	6	1
5	3	-2

(d)

4	-3
2	3
-5	6

Exercise 5.2   Investigate the saddle-points, if any, of the following and give the value of the game where appropriate.

(a)

	b1	b2	b3	b4
a1	-6	13	12	5
a2	8	1	9	0
a3	0	-5	13	14
a4	3	18	10	9

(b)

	b1	b2	b3	b4
a1	2	-1	-1	0
a2	0	1	-2	0
a3	-2	-3	-3	-2
a4	-1	2	-1	-1
a5	-2	-3	-4	-2

(c)

	b1	b2	b3	b4	b5
a1	3	-1	0	2	5
a2	6	4	5	4	8
a3	5	1	7	-2	2
a4	9	4	10	4	4
a5	2	3	-1	0	1
a6	7	-2	3	1	5

(d)

	b1	b2	b3
a1	3	1	5
a2	5	4	-3
a3	2	0	1

11.6 EXPECTATION = EXPECTED PAY-OFF [= MEAN WINNINGS PER GAME]

If a player takes part in a game with three possible outcomes as given in the table, what is his expectation?

Winnings	£2	£3	–£5
Probability	$\dfrac{1}{2}$	$\dfrac{1}{3}$	$\dfrac{1}{6}$

In the long run he would expect to win £2 half the times he played, £3 on one third of the occasions and lose £5 the rest of the time.

A series of 6000 games might produce results for him of

$$\text{total pay-out} = £2 \times 3000 + £3 \times 2000 - £5 \times 1000$$

expectation = average pay-out per game

$$= \frac{£2 \times 3000}{6000} + \frac{£3 \times 2000}{6000} - \frac{£5 \times 1000}{6000}$$

$$= £2 \times \frac{1}{2} + £3 \times \frac{1}{3} - £5 \times \frac{1}{4}$$

$$= £2 \times (P \text{ winning } £2) + £3 \times (P \text{ winning } £3) - £5 \times (P \text{ winning } -£5)$$

$$\left[= \Sigma i\, P(i) \right]$$

In this case, expectation $\quad = 1 + 1 - \dfrac{5}{6}$

$$= £1\frac{1}{6} \text{ (per game)}.$$

11.7 MINIMAX MIXED STRATEGY

Consider the pay-off matrix for a two-person game given below.

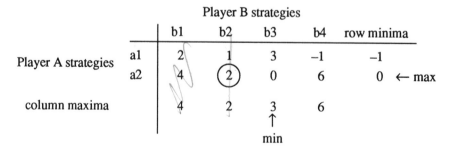

		Player B strategies				
		b1	b2	b3	b4	row minima
Player A strategies	a1	2	1	3	–1	–1
	a2	4	2	0	6	0 ← max
column maxima		4	2	3	6	

min

It is not profitable for A to switch strategies as his pay-off would decrease from 2 to 1. This assumes that B plays b2, but there is an unstable solution here as B would gain by switching to b3 where his loss is zero.

Since B is likely to change his approach, A must be prepared to do so as well and must adopt a mixed strategy, i.e. playing a1 some of the time and a2 some of the time.

Let $p = P$ (A plays a1), then $1 - p = P$ (A plays a2).

If B plays b1, the expectation for A is E (A plays b1) $= 2p + 4(1-p) = 4 - 2p$.

This can be shown graphically (Fig. 11.1).

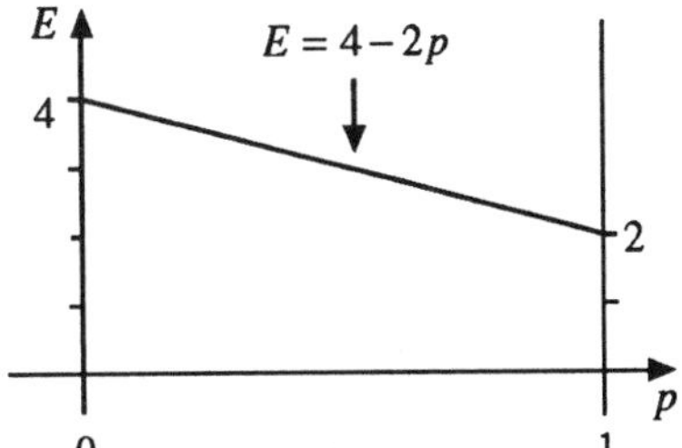

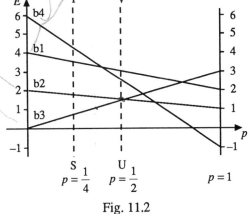

Fig. 11.1

When $p = 1$, A plays a1 all the time and against b1 wins 4. When $p = 0$, A plays a2 all the time and against b1 wins 2.

$$E \text{ (A plays b1)} = 2p + 4(1-p) = 4 - 2p$$
$$E \text{ (A plays b2)} = 1p + 2(1-p) = 2 - p$$
$$E \text{ (A plays b3)} = 3p + 0(1-p) = 3p$$
$$E \text{ (A plays b4)} = -1p + 6(1-p) = 6 - 7p$$

Putting all these on the same diagram gives the array in Fig. 11.2.

Fig. 11.2

If A chooses to play strategy a1 with probability $\frac{1}{4}$ (and a2 with probability $\frac{3}{4}$) then the vertical line ST gives the expected pay-off under each of the four options available to B.

Best play for B would be strategy b3 when the expected pay-out for A is $3p = \frac{3}{4}$ as long as $p = \frac{1}{4}$.

A should select the mixed strategy (i.e. value of p) which maximises this minimum payout.

This is given by the line UV, where $p = \dfrac{1}{2}$, and his expectation is

$$E \text{ (A plays b1)} = 3, \ E \text{ (A plays b2)} = \frac{3}{2}, \ E \text{ (A plays b3)} = \frac{3}{2}, \ E \text{ (A plays b4)} = \frac{5}{2}.$$

Hence A should play a1 and a2 on the spin of a fair coin and the **value** of the game to him is $\dfrac{3}{2}$.

$$\text{MAXIMIN} \leq \text{VALUE} \leq \text{MINIMAX}$$

What about B's best play?

Since there are four strategies to 'pick and mix', there are several probabilities to be assigned. As you now have several variables determining the expectation you cannot draw a simple graph as was possible for A, so you might be in the realms of linear programming.

You might have noticed that B will never play b1 as his losses here are always greater than those from playing b2. The matrix can be reduced to

Player B

		b2	b3	b4
	a1	1	3	−1
Player A	a2	2	0	6

but with no further simplification possible you are still unable to use a graphical method of solution. The linear programming problem will, though, have been made easier.

The irrelevance of b1 can also be seen from your graph. Since its line was never the lowest it was never going to come into any considerations.

Looking back at the matrix and graph, you can see that the lines come from joining 4 to 2, to 1, etc. on the $p = 0$ and $p = 1$ lines, and these numbers are the bottom and top rows of the array.

11.8 GAMES WITH KNOWN VALUES

The game given by the array has a value of $\dfrac{3}{2}$ to A and, therefore, of $-\dfrac{3}{2}$ to B.

Player B

		b2	b3	b4
	a1	1	3	−1
Player A	a2	2	0	6

How should B play to achieve the value $-\dfrac{3}{2}$, which is the best he can do?

$p = P$ (B plays b2), $q = P$ (B plays b3), $1 - p - q = P$ (B plays b4)

$$E \text{ (B plays a1)} \;=\; p + 3q - (1 - p - q) \;=\; \frac{3}{2} \qquad\qquad *$$

$$E \text{ (B plays a2)} \;=\; 2p + 6(1 - p - q) \;=\; \frac{3}{2} \qquad\qquad *$$

$$\Rightarrow \qquad 2p + 4q - 1 \;=\; \frac{3}{2}$$

$$\text{and} \qquad 6 - 4p - 6q \;=\; \frac{3}{2}$$

$$\Rightarrow \qquad 2p + 4q \;=\; \frac{5}{2} \qquad\qquad (1)$$

$$\text{and} \qquad 4p + 6q \;=\; \frac{9}{2} \qquad\qquad (2)$$

$$(1) - (2) \div 2 \qquad \Rightarrow \qquad q \;=\; \frac{1}{4}$$

$$(1) \qquad \Rightarrow \; 2p + 1 = \frac{5}{2} \;\Rightarrow\; p = \frac{3}{4}$$

Therefore B should play b2 three quarters of the time and b3 the remaining quarter.

Note that the equations * have $\dfrac{3}{2}$ and not $-\dfrac{3}{2}$ on the right hand side. This is because the values on the left are from the matrix and so are B's losses, which are to give an overall loss of $\dfrac{3}{2}$.

Exercise 8.1   Find the best mixed strategies for each player and the value of the game to each in the two tables (a) and (b) below.

(a)

	Player B	
	b1	b2
Player A a1	1	4
a2	2	-3

(b)

	Player B	
	b1	b2
Player A a1	1	4
a2	4	2

Exercise 8.2 **(Investigation)**

Use a graphical method to find the best mixed strategy for the machine and the value of the game. By using the known value of the game, or otherwise, find the player's best mixed strategy.

	Machine	
	m1	m2
Player p1	6	-2
p2	-4	3
p3	-2	-1

11.9 USING LINEAR PROGRAMMING

Consider the pay-off matrix for a 3×2 game given below.

		b1	b2	row minima	
	a1	11	3	3	← max
Player A	a2	1	8	1	
strategies	a3	3	4	3	← max
column maxima		11	8		

Player B
strategies

↑
min

Look first to see if the matrix can be reduced. Neither comparing rows nor comparing columns will help make the problem any smaller.

The initial investigation leads you to look for a minimax mixed strategy as the solution cell a1 b2 is unstable.

$$E \text{ (A plays b1)} = 11p_1 + p_2 + 3p_3$$

and $$E \text{ (A plays b2)} = 3p_1 + 8p_2 + 4p_3$$

where p_1 is the probability that A plays strategy a1, etc.

Looking back at the graph in section 11.7, you can see that the aim was to find the tallest vertical line that was below, or touching, the sloping expectation lines. This height was the value of the game. Similarly in the linear programming situation you seek a value as large as possible that is less than or equal to the expectations.

If the value of the game is v, then the problem is to maximise the objective function

$$P = v$$

subject to $$11p_1 + p_2 + 3p_3 \geq v$$
$$3p_1 + 8p_2 + 4p_3 \geq v$$
$$p_1 + p_2 + p_3 = 1$$

Since $p_1 + p_2 + p_3 = 1$ it is certainly true that $p_1 + p_2 + p_3 \leq 1$.
Using this inequality instead of the equation allows the introduction of an extra slack variable. This makes it simple to write a basic feasible solution and provides an instant canonical form. The slack variable introduced (z) will certainly turn out to be zero.

Maximise $P = v$

subject to $$v - 11p_1 - p_2 - 3p_3 \leq 0$$
$$v - 3p_1 - 8p_2 - 4p_3 \leq 0$$
$$p_1 + p_2 + p_3 \leq 1$$

Introducing slack variables,

$$v - 11\,p_1 \;-\; p_2 \;-3\,p_3 \;+\; x \qquad\qquad = 0$$
$$v - \;\; 3\,p_1 \;-8\,p_2 \;-4\,p_3 \qquad\quad +\; y \qquad = 0$$
$$p_1 \;+\; p_2 \;+\; p_3 \qquad\qquad\qquad +\; z \;= 1$$

A basic feasible solution is $x = 0,\, y = 0,\, z = 1,\, p_1 = p_2 = p_3 = 0,\, v = 0$ which gives the objective function a value of zero. (This is only to be expected when no game is being played!)

The initial tableau

	basis	value	v	p_1	p_2	p_3	x	y	z	check
(1)	x	0	1*	−11	−1	−3	1	•	•	−13
(2)	y	0	1	−3	−8	−4	•	1	•	−13
(3)	z	1	•	1	1	1	•	•	1	4
(4)	P	0	1	•	•	•	•	•	•	1

↑
v could replace either x or y at this stage

(5)	v	0	1	−11	−1	−3	1	•	•	−13
(6)	y	0	•	8*	−7	−1	−1	1	•	0
(7)	z	1	•	1	1	1	•	•	1	4
(8)	P	0	•	11	1	3	−1	•	•	14

↑

$$(6) = (2) - (1) \qquad (7) = (3) \qquad (5) = (1) \qquad (8) = (4) - (1)$$

(9)	v	0	1	•	$-\frac{85}{8}$	$-\frac{35}{8}$	$-\frac{3}{8}$	$\frac{11}{8}$	•	−13
(10)	p_1	0	•	1	$-\frac{7}{8}$	$-\frac{1}{8}$	$-\frac{1}{8}$	$\frac{1}{8}$	•	0
(11)	z	1	•	•	$\frac{15}{8}$ *	$\frac{9}{8}$	$\frac{1}{8}$	$-\frac{1}{8}$	1	4
(12)	P	0	•	•	$\frac{85}{8}$	$\frac{35}{8}$	$\frac{3}{8}$	$-\frac{11}{8}$	•	14

↑

$$(10) = (6) \div 8 \qquad (9) = (5) + 11(1) \qquad (11) = (7) - (10) \qquad (12) = (8) - 11(10)$$

(13)	v	$\frac{17}{3}$	1	•	•	2	$\frac{1}{3}$	$\frac{2}{3}$	$\frac{17}{3}$	$\frac{29}{3}$
(14)	p_1	$\frac{7}{15}$	•	1	•	$\frac{2}{5}$	$-\frac{1}{15}$	$\frac{1}{15}$	$\frac{7}{15}$	$\frac{28}{15}$
(15)	p_2	$\frac{8}{15}$	•	•	1	$\frac{3}{5}$	$\frac{1}{15}$	$-\frac{1}{15}$	$\frac{8}{15}$	$\frac{32}{15}$
(16)	P	$-\frac{17}{3}$	•	•	•	−2	$-\frac{1}{3}$	$-\frac{2}{3}$	$-\frac{17}{3}$	$-\frac{26}{3}$

$$(15) = \tfrac{8}{15}(11) \qquad (14) = (10) + \tfrac{7}{15}(11) \qquad (13) = (9) + \tfrac{85}{15}(11)$$

$$(16) = (12) - \tfrac{17}{3}(11) \qquad\qquad\qquad\qquad = (9) + \tfrac{17}{3}(11)$$

$$\Rightarrow \text{value} = \tfrac{17}{3} \quad \text{when } p_1 = \tfrac{7}{15}, \ p_2 = \tfrac{8}{15}, \ p_3 = 0.$$

The game with pay-off matrix shown here presents a problem.
In linear programming terms you are going to maximise $P = v$

$$\begin{array}{c|cc} & 6 & -2 \\ \hline & -4 & 3 \\ & -2 & -1 \end{array}$$

subject to
$$6p_1 - 4p_2 - 2p_3 \geq v$$
$$-2p_1 + 3p_2 - p_3 \geq v$$
$$p_1 + p_2 + p_3 \leq 1$$

Introducing slack variables

$$v - 6p_1 + 4p_2 + 2p_3 + x = 0$$

$$v + 2p_1 - 3p_2 + p_3 \quad + y = 0$$

$$p_1 + p_2 + p_3 + \quad + z = 1$$

Since linear programming is designed to deal with problems based on variables that do not take on negative values, this matrix could cause problems. The existence of minus signs in the original array means that the value of the game, v, could be negative.

To avoid any difficulties add four to each element of the matrix, giving this matrix. Find the value and remember to subtract four at the end.

$$\begin{array}{cc} 10 & 2 \\ 0 & 7 \\ 2 & 3 \end{array}$$

Exercise 9.1   Write the following zero-sum games as linear programming problems to find player A's best mixed strategies.

(a)

$$\begin{array}{c|cc} & \multicolumn{2}{c}{B} \\ \hline A & 1 & 4 \\ & 4 & 2 \end{array}$$

(b)

$$\begin{array}{c|cc} & \multicolumn{2}{c}{B} \\ \hline A & 1 & 4 \\ & 2 & -3 \end{array}$$

(Check your answers with Exercise 8.1 in section 11.8.) Find the values of the games to A.

Exercise 9.2   Find the best mixed strategy for A and the value of the game with pay-off matrix

$$\begin{array}{c|cc} & \multicolumn{2}{c}{B} \\ \hline & 14 & 16 \\ A & 19 & 7 \\ & 22 & -3 \end{array}$$

11.10 MISCELLANEOUS EXERCISES

Exercise 10.1   Decide which of the following have stable solutions and find the values of the games where appropriate.

(a)

	B		
	6	4	1
A	4	5	-2
	-1	1	3

(b)

	B		
	5	3	7
A	6	1	0
	-1	2	8

(c)

	B			
	14	2	2	7
A	6	2	2	9
	15	-1	1	0

Exercise 10.2
(a)   Reduce the following pay-off matrix as far as possible, explaining your method.

	B			
	2	3	5	9
A	0	14	2	1
	8	6	7	10

(b)   Formulate the problem of finding B's best mixed strategy from your reduced matrix by setting up suitable inequalities and/or equations to be solved by linear programming. Do not solve the problem.

Exercise 10.3   Reduce the matrix given here as far as possible. How do you know there is no stable solution? Find the best mixed strategy for each player and the value of the game.

	B		
	2	1	5
A	-2	7	4

11.11   NOTES
Expanding square (for $n \times 2$ games)

Example

Consider the pay-off matrix for a 3×2 game which was tackled by linear programming in section 11.9.

	b1	b2
a1	11	3
a2	1	8
a3	3	4

To find A's best strategy, plot the returns to him as points and join them to give a region.

Expand a square with bottom left corner at the origin, and two other vertices on the axes, until it is about to leave the region. It hits the line joining a1 to a2 indicating that only these will be involved in A's mixed strategy, i.e. $p_3 = 0$. To find the values of p_1 and p_2 look at the ratio in which the line from a1 to a2 has been divided.

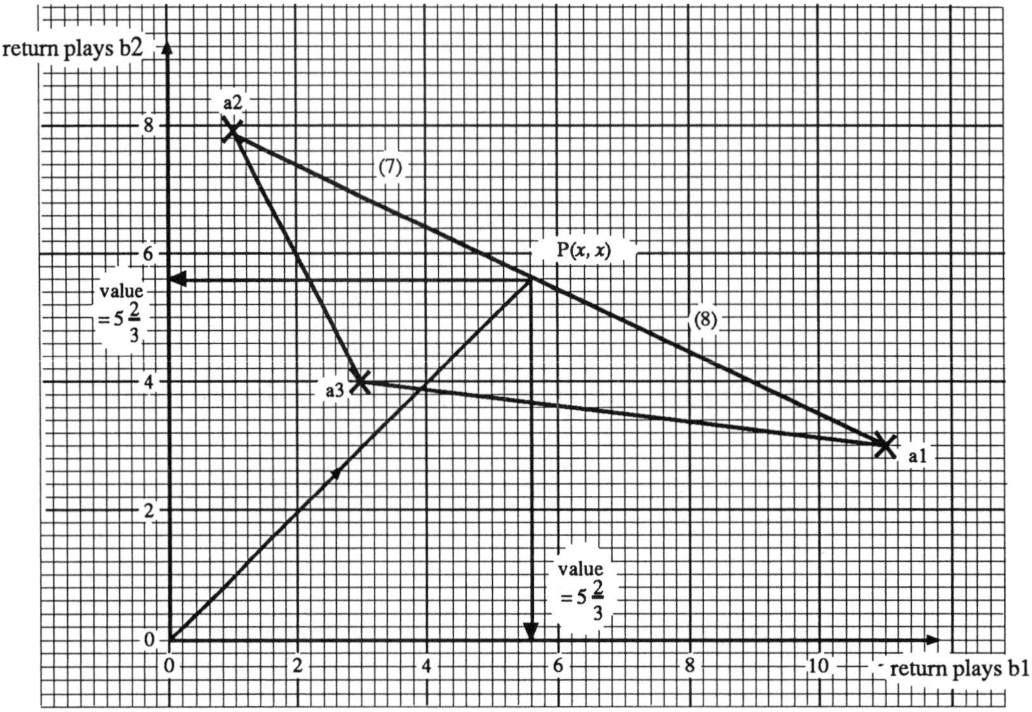

Fig. 11.3

gradient a2P = gradient a1P

$$\Rightarrow \qquad \frac{x-8}{x-1} = \frac{x-3}{x-11}$$

$$\Rightarrow \qquad (x-8)(x-11) = (x-3)(x-1)$$

$$\Rightarrow \qquad x^2 - 19x + 88 = x^2 - 4x + 3$$

$$\Rightarrow \qquad 85 = 15x$$

$$\Rightarrow \qquad x = 5\frac{2}{3}$$

$$5\frac{2}{3} - 1 : 11 - 5\frac{2}{3} = 17 - 3 : 33 - 17$$

$$= 14 : 16$$

$$= 7 : 8$$

$$\Rightarrow \qquad p_1 = \frac{7}{15} \qquad p_2 = \frac{8}{15}$$

You can see from the example that at most two 'p's are non-zero in these cases. As with linear programming the entries in the matrix must be non-negative.

For example, the pay-off matrix

	b1	b2
a1	2	1
a2	5	2
a3	4	3

gives a diagram where the top right corner of the square never hits the region (Fig. 11.4).

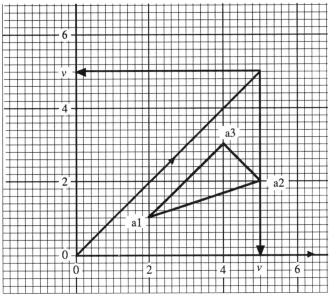

Fig. 11.4

This is because the x-coordinate is always greater than the y-coordinate, i.e. B's losses under b1 are always more than under b2, so the matrix can be reduced. Look for **reduction** first!

If the top right corner leaves through a vertex, the solution is stable. Look for **stability** second!

Never use a sledgehammer to crack a walnut!

12

Recurrence relations

12.1 WHAT IS A RECURRENCE RELATION?

A **recurrence relation** is literally a relation that allows you to run on and on through the terms of a sequence. For example,

$$u_{n+1} = 3u_n + 2 \quad \text{and} \quad u_0 = 1$$

says that any term in a sequence is found by multiplying the previous term by three and adding two and the first term is one. This gives a way of running through the terms of a sequence, so

$$n = 0 \qquad u_1 = 3u_0 + 2 = 3 \times 1 + 2 = 5$$
$$n = 1 \qquad u_2 = 3u_1 + 2 = 3 \times 5 + 2 = 17$$
$$n = 2 \qquad u_3 = 3u_2 + 2 = 3 \times 17 + 2 = 53, \text{ etc.}$$

But what is u_{100}, the hundredth term of the sequence? Do you need to work out all the intermediate terms or can you find a formula that generates the numbers 1, 5, 17, 53, ... ?

Since each term comes from the one before it by multiplying by three, as part of the process, looking at powers of three would be a sensible place to start.

n	0		1		2		3
u_n	1	$\xrightarrow{4}$	5	$\xrightarrow{12}$	17	$\xrightarrow{36}$	53
powers of 3: 3^n	1	$\xrightarrow{2}$	3	$\xrightarrow{6}$	9	$\xrightarrow{18}$	27

The u_n sequence goes up twice as fast as the powers of 3 so multiply the bottom row by two.

n	0	1	2	3
u_n	1	5	17	53
2×3^n	2	6	18	54

Now is it possible to see the connection between the two sequences, that is,

$$u_n = 2 \times 3^n - 1$$

so $\qquad u_{100} = 2 \times 3^{100} - 1 \qquad \left(\approx 10^{48} \right)$

12.2 INVESTIGATIONS
Use the ideas of section 12.1 to produce a formula and the hundredth term for each of the following.

(a) $u_{n+1} = 2u_n + 3$ $\qquad\qquad u_0 = 1$

(b) $u_{n+1} = 3u_n - 2$ $\qquad\qquad u_0 = 1$

(c) $u_{n+1} = 0.5u_n + 1$ $\qquad\qquad u_0 = 4$

(d) $u_{n+1} = 2u_n + n$ $\qquad\qquad u_0 = 1$

12.3 FIRST ORDER LINEAR EQUATIONS
An equation is of the **first order** (or is of order one) if the difference between the highest and lowest subscripts is one.

So $\qquad u_{n+1} = 2u_n + 3 \qquad$ is order one

and $\qquad u_{n+2} = 2u_n + 3 \qquad$ is order two.

Each of the equations above is linear as the 'u's only appear raised to the power one, and only consist of constant multiples of them added or subtracted.

So

$$u_{n+1} - au_n^2 = b$$

$$u_{n+1} \times u_n = b$$

$$u_{n+1} - a2^{u_n} = b$$

are not linear. (The second equation has two 'u' terms multiplied so effectively has a 'u' squared on the left.)

Homogeneous equations
An equation is **homogeneous** if all its terms contain 'u's to the same power.

Example 1		$u_{n+1} = u_n$	and	$u_0 = 3$
Solution	$n = 0$	$u_1 = u_0$		$u_1 = 3$
	$n = 1$	$u_2 = u_1$		$u_2 = 3$
	$n = 2$	$u_3 = u_2$		$u_3 = 3$

So $\qquad u_n = 3$

In general,

$$u_{n+1} = u_n \quad \text{and} \quad u_0 = K \quad \Rightarrow \quad u_n = K$$

Example 2 $u_{n+1} = 2u_n$ and $u_0 = 3$

Solution

$n = 0$	$u_1 = 2u_0 = 2 \times 3$	$u_1 = 6$	$(= 3 \times 2)$
$n = 1$	$u_2 = 2u_1 = 2 \times 6$	$u_2 = 12$	$(= 3 \times 2 \times 2)$
$n = 2$	$u_3 = 2u_2 = 2 \times 12$	$u_3 = 24$	$(= 3 \times 2 \times 2 \times 2)$

So $u_n = 3 \times 2^n$

In general,

$$u_{n+1} = au_n \quad \text{and} \quad u_0 = K \quad \Rightarrow \quad u_n = K \times a^n$$

Example 1 is just Example 2 with $a = 1$.

Inhomogeneous equations

Equations of the form $u_n = au_{n-1} + c$ are called **inhomogeneous** (or **non-homogeneous**) equations.

Example $u_{n+1} = 3u_n + 5$ and $u_0 = 2$

Solution $u_{n+1} = 3u_n$ has a solution $u_n = K \times 3^n$

so for $u_{n+1} = 3u_n + 5$ try a solution $u_n = A \times 3^n + B$

To find two constants you need two facts. You have $u_0 = 2$ and $u_1 = 3 \times u_0 + 5 = 11$.

$$u_n = A \times 3^n + B \qquad \begin{array}{ll} n = 0 \Rightarrow & u_0 = A \times 1 + B = 2 \qquad (1) \\ n = 1 \Rightarrow & u_1 = A \times 3 + B = 11 \qquad (2) \end{array}$$

Solve the simultaneous equations by subtracting the first from the second.

$$(2) - (1) \quad \Rightarrow \quad 2A = 9 \quad \Rightarrow \quad A = \frac{9}{2}$$

$$(1) \quad \Rightarrow \quad A + B = 2 \quad \Rightarrow \quad B = -\frac{5}{2}$$

$$u_n = \frac{9}{2} \times 3^n - \frac{5}{2}$$

which can be checked by using it to produce the term u_2. The recurrence relation gives

$$u_2 = 3 \times u_1 + 5 = 3 \times 11 + 5 = 38$$

which agrees with $u_n = \frac{9}{2} \times 3^2 - \frac{5}{2} = \frac{81}{2} - \frac{5}{2} = 38$

In general,

$$u_{n+1} = au_n + b \quad \text{and} \quad u_0 + K \quad \Rightarrow \quad u_n = A \times a^n + B$$

Note The two equations $u_n = 3u_{n-1} + 7$ and $u_{n+1} = 3u_n + 7$ give exactly the same information. Each equation says that a term is obtained from the preceding one by multiplying it by three and adding seven.

Exercise 3.1   Use the results of section 12.3 to find a general formula for u_n for each of the following:

(a) $u_{n+1} = 7u_n + 6$ $u_0 = 0$

(b) $u_n = 5u_{n-1} + 2$ $u_0 = 3$

(c) $u_{n+1} - 0.8u_n = 0$ $u_0 = 10$

(d) $u_{n+1} = 7 - u_n$ $u_0 = 6$

Exercise 3.2   Ashoke puts £100 into a savings account which pays 8% interest on 1st January every year. Explain why $x_{n+1} = 1.08x_n + 100$ and $x_0 = 100$ where x_n is the amount of money in the account at midnight on 1 January after n years. Find a formula for x_n and use it to find how much he will have after his 20th deposit.

Exercise 3.3   Transfer all the discs from post A to post C moving only one disc at a time, and with no disc ever placed on top of a smaller one.

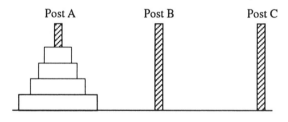

Post A   Post B   Post C

Fig. 12.1

If m_n is the smallest number of moves needed for n discs, explain why $m_4 = 2m_3 + 1$.

Write down a recurrence relation for m_{n+1} in terms of m_n and solve it to find the number of moves needed for n discs.

12.4 COMPLEMENTARY FUNCTIONS AND PARTICULAR SOLUTIONS
The general solution of $u_{n+1} = au_n + b$ and $u_0 = K$

is $u_n = A \times a^n + B$.

The $A \times a^n$ part is called the **complementary function** and you know that it satisfies $u_{n+1} = au_n$.

The 'B' term has been added to produce the 'b' required in the recurrence relation and is called the **particular solution**.

Example

$$u_{n+1} = 15 - 2u_n \quad \text{and} \quad u_0 = 6$$

Solution

$$u_n = A(-2)^n + B$$
$$\Rightarrow \quad u_{n+1} = A(-2)^{n+1} + B$$

Substituting the particular solution, B, into the recurrence equation gives

$$\Rightarrow \quad B = 15 - 2B$$
$$\Rightarrow \quad 3B = 15$$
$$\Rightarrow \quad B = 5$$

$$n = 0 \quad \Rightarrow \quad u_0 = A + B$$
$$\Rightarrow \quad 6 = A + 5$$
$$\Rightarrow \quad A = 1$$

so $\qquad u_n = (-2)^n + 5$

Substituting the whole general solution in would have given

$$A(-2)^{n+1} + B = 15 - 2\left[A(-2)^n + B \right]$$
$$\Rightarrow \quad A(-2)^{n+1} + B = 15 + A(-2)^{n+1} - 2B$$
$$\Rightarrow \quad B = 15 - 2B \qquad \qquad \text{i.e. the same.}$$

The A vanished, which is not really surprising as it had no part in providing the $b = 15$ anyway.

The same method as was used in section 12.3 could have been employed here by forming a pair of simultaneous equations from u_0 and u_1. The main advantage of this second plan is that u_0 is not required to find B. The 'b' alone determines 'B' and then u_0 fixes the value of A.

Example

$$u_{n+1} = 3u_n - 4n \quad \text{and} \quad u_0 = 4$$

Solution

Try $\qquad u_n = A3^n + Bn + C \quad$ as a general solution.

Method 1 (from section 12.3)

$$n = 0 \qquad \qquad u_1 = 3u_0 - 4 \times 0 = 12 \quad \text{from the recurrence relation and}$$
$$n = 1 \qquad \qquad u_2 = 3u_1 - 4 \times 1 = 32$$

General solution,

$$n = 0 \quad \Rightarrow \quad u_0 = A \quad\quad + C = 4 \tag{1}$$

$$n = 1 \quad \Rightarrow \quad u_1 = 3A + B \;+ C = 12 \tag{2}$$

$$n = 2 \quad \Rightarrow \quad u_2 = 9A + 2B + C = 32 \tag{3}$$

gives three simultaneous equations to solve.

$$(3) - 2(2) \quad \Rightarrow \quad 3A - C = 8$$
$$(1) \quad \Rightarrow \quad A + C = 4$$
$$\text{ADD} \quad \Rightarrow \quad 4A = 12$$
$$A = 3$$

$$(1) \quad \Rightarrow \quad 3 + C = 4 \quad \Rightarrow \quad C = 1$$
$$(2) \quad \Rightarrow \quad 9 + B + 1 = 12 \quad \Rightarrow \quad B = 2$$
$$\Rightarrow \quad u_n = 3 \times 3^n + 2n + 1$$
$$= 3^{n+1} + 2n + 1$$

Method 2 (from section 12.4)

$$u_{n+1} = A3^{n+1} + B(n+1) + C$$

Substitute the particular solution into the recurrence relation.

$$\Rightarrow \quad B(n+1) + C = 3(Bn + C) - 4n$$
$$Bn + B + C = 3Bn - 4n + 3C$$

number of n on each side: $\quad B = 3B - 4 \Rightarrow B = 2$

constant on each side: $\quad B + C = 3C \quad\quad \Rightarrow C = 1$

so the general solution is $\quad u_n = A \times 3^n + 2n + 1$

$$u_0: \quad 4 = A \times 1 + 2 \times 0 + 1$$
$$\Rightarrow \quad A = 3$$
$$\Rightarrow \quad u_n = 3 \times 3^n + 2n + 1$$
$$\Rightarrow \quad u_n = 3^{n+1} + 2n + 1$$

In general,

$$u_{n+1} = au_n + bn + c \quad\quad \Rightarrow \quad u_n = A \times a^n + Bn + C$$
$$u_{n+1} = au_n + bn^2 + cn + d \quad \Rightarrow \quad u_n = A \times a^n + Bn^2 + Cn + D$$
$$\text{etc.}$$

Note   It is **not** true that $u_{n+1} = au_n + bn$ has general solution $u_n = A \times a^n + Bn$, as can be seen from the last example. The fact that 'c' is zero does not mean that C is also zero.

Exercise 4.1   Solve the following, leaving one undetermined constant in each.

(a) $u_{n+1} = \dfrac{u_n}{3} + 1$ (b) $u_{n+1} = 5u_n - 4$ (c) $u_{n+1} = 7u_n + 12n$

Exercise 4.2   Find the solution of $u_{n+1} = 2u_n + n$ and $u_0 = 0$.

Exercise 4.3   Show that the general solution of $u_{n+1} - 2u_n = n^2$ and $u_0 = 0$ is $u_n = 3 \times 2^n - n^2 - 2n - 3$.

12.5 INHOMOGENEOUS $u_{n+1} = au_n + \lambda^n$
(i) $a \neq \lambda$

Example
$$u_{n+1} = 2u_n + 3^n \qquad u_0 = 2$$

Solution

Try $u_n = A2^n + B3^n$ as general solution.

$$n = 0 \qquad \Rightarrow u_1 = 2u_0 + 3^0 = 2 \times 2 + 1 = 5$$

General solution

$$\Rightarrow \quad u_0 = A \times 1 + B \times 1 = 2 \qquad \Rightarrow \qquad A + B = 2 \qquad (1)$$

and $u_1 = A \times 2 + B \times 3 = 5 \qquad \Rightarrow \qquad 2A + 3B = 5 \qquad (2)$

$(2) - 2(1) \Rightarrow \qquad B = 1 \qquad \Rightarrow \qquad A = 1$

$$\Rightarrow \qquad u_n = 2^n + 3^n$$

(ii) $a = \lambda$

Example
$$u_{n+1} = 2u_n + 2^n \qquad u_0 = 1$$

Solution

There is no point trying $B \times 2^n$ as a particular solution to provide the 2^n in the recurrence relation as this would give

$$u_n = A \times 2^n + B \times 2^n = (A + B) \times 2^n$$

$$= \text{constant} \times 2^n \text{ is the complementary function.}$$

Try $u_n = A2^n + Bn2^n$

$n = 0$ $\Rightarrow$ $u_1 = 2 \times u_0 + 2^0 = 2 \times 1 + 1 = 3$

general solution

$\Rightarrow$ $u_0 = A \times 1 + B \times 0 = 1$ $\Rightarrow$ $A = 1$

$u_1 = A \times 2 + B \times 2 = 3$ $\Rightarrow$ $2A + 2B = 3$

$\Rightarrow$ $B = \dfrac{1}{2}$

$\Rightarrow$ $u_n = 2^n + \dfrac{n}{2} \times 2^n$

12.6 INHOMOGENEOUS: $a = 1$
Example

$u_{n+1} = u_n + n$ $u_0 = 1$

Solution

A complementary function $A \times 1^n = A$ together with a particular solution $Bn + C$ will **not** provide a general solution.

If you try $u_n = A + Bn + C$

$\Rightarrow$ $u_{n+1} = A + B(n+1) + C$

Then $A + B(n+1) + C = A + Bn + C + n$ from the recurrence equation

$\Rightarrow$ $Bn + A + B + C = Bn + n + A + C$

no. of n on each side $B = B + 1$ $\Rightarrow$ $B = \infty$

constant on each side: $A + B + C = A + C$ $\Rightarrow$ $B = 0$!!

Try $u_n = An^2 + Bn + C$

$u_{n+1} = u_n + n,$
$n = 0$ $\Rightarrow$ $u_1 = u_0 + 0 = 1$

$n = 1$ $\Rightarrow$ $u_2 = u_1 + 1 = 2$

$u_n = An^2 + Bn + C,$
$n = 0$ $\Rightarrow$ $u_0 = C = 1$ (1)

$n = 1$ $\Rightarrow$ $u_1 = A + B + C = 1$ (2)

$n = 2$ $\Rightarrow$ $u_2 = 4A + 2B + C = 2$ (3)

$(2) - (1)$ $\Rightarrow$ $A + B = 0$

$(3) - (2)$ $\Rightarrow$ $3A + B = 1$

$$\Rightarrow \quad A = \frac{1}{2}, \quad B = -\frac{1}{2}$$

$$u_n = \frac{1}{2}n^2 - \frac{1}{2}n + 1$$

In general,

$$u_{n+1} = u_n + \text{ polynomial of degree (highest power) } N$$

$$\Rightarrow u_n = \text{ polynomial of degree } N + 1$$

Example

$$u_{n+1} = u_n + 3^n \qquad u_0 = 0$$

Solution

Try $\qquad u_n = A \times 1^n + B \times 3^n = A + B \times 3^n \quad$ (i.e. the same as the case $a \neq 1$)

$u_{n+1} = u_n + 3^n,$

$\quad n = 0 \qquad \Rightarrow \qquad u_1 = u_0 + 1 = 1$

$u_n = A + B \times 3^n,$

$\quad n = 0 \qquad \Rightarrow \qquad u_0 = A + B = 0 \qquad\qquad\qquad\qquad (1)$

$\quad n = 1 \qquad \Rightarrow \qquad u_1 = A + 3B = 1 \qquad\qquad\qquad\qquad (2)$

$(2) - (1) \quad \Rightarrow \qquad 2B = 1 \Rightarrow B = \frac{1}{2}, \ A = -\frac{1}{2}$

$$\Rightarrow \qquad u_n = \frac{1}{2} \times 3^n - \frac{1}{2}$$

Exercise 6.1   Solve the following recurrence relations.

(a) $\ u_{n+1} - 2u_n = 5^n \qquad\qquad u_0 = 1$

(b) $\ u_{n+1} + 3u_n = 7^n \qquad\qquad u_0 = 1$

(c) $\ u_{n+1} - 3u_n = 3^n \qquad\qquad u_0 = 2$

(d) $\ u_{n+1} - u_n = 2n + 1 \qquad\quad u_0 = 0$

(e) $\ u_{n+1} - u_n = 2 \qquad\qquad\quad u_0 = 0$

Exercise 6.2   Let u_n be the number of diagonals of a regular n-gon. When the number of vertices is increased by one the number of diagonals increases by $n - 2 + 1 = n - 1$ as the new vertex, A, can form $n - 2$ diagonals with existing vertices (not with 1 or 2) and the edge 1-2 becomes a diagonal. Write down a recurrence relation connecting u_{n+1} and u_n. Solve your equation to give a formula for u_n.

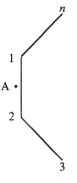

Fig. 12.2

(**Note** u_0, u_1, u_2 do not exist.)

Exercise 6.3   Solve $u_{n+1} - u_n = 5^n$ $u_0 = 1$

12.7 SECOND ORDER LINEAR
Homogeneous
(i) Distinct real roots

Example

$$u_{n+2} - 5u_{n+1} + 6u_n = 0 \qquad u_0 = 3, \ u_1 = 7$$

Solution

Try a general solution:

$$u_n = Ax^n \qquad \Rightarrow \qquad u_{n+1} = Ax^{n+1}, \qquad u_{n+2} = Ax^{n+2}$$

$$u_{n+2} - 5u_{n+1} + 6u_n = 0 \quad \Rightarrow \qquad Ax^{n+2} - 5Ax^{n+1} + 6Ax^n = 0$$

$$\Rightarrow \qquad x^{n+2} - 5x^{n+1} + 6x^n = 0$$

It is permissible to divide by A as it cannot be zero. If $A = 0 \Rightarrow u_n = 0$ which will not satisfy $u_0 = 3, \ u_1 = 7$.

Dividing by $x_n \Rightarrow x^2 - 5x + 6 = 0$, which is called the **auxiliary equation**. Compare it with the original recurrence relation at the start of this example.

Again it is possible to divide by x^n as x is not zero. If it were, then $u_n = 0$ for all values of n.

$$(x - 2)(x - 3) = 0$$

$$\Rightarrow \qquad x = 2 \text{ or } 3$$

General solution

$$\Rightarrow \qquad u_n = A2^n + B3^n$$

$$n = 0 \qquad \Rightarrow \qquad u_0 = A + B = 3 \tag{1}$$

$$n = 1 \qquad \Rightarrow \qquad u_1 = 2A + 3B = 7 \tag{2}$$

$$(2) - 2(1) \qquad \Rightarrow \qquad B = 1$$

$$(1) \qquad \Rightarrow \qquad A = 2$$

$$\Rightarrow \quad u_n = 2 \times 2^n + 1 \times 3^n$$

$$= 2^{n+1} + 3^n$$

Check: $u_2 - 5u_1 + 6u_0 = 0$

$u_1 = 7, \ u_0 = 3 \quad \Rightarrow \qquad u_2 - 5 \times 7 + 6 \times 3 = 0$

$$\Rightarrow \qquad u_2 = 35 - 18 = 17$$

$u_n = 2^{n+1} + 3^n, \ n = 2 \quad \Rightarrow \qquad u_2 = 2^3 + 3^2$

$$= 8 + 9 = 17$$

In practice, the amount of working that needs to be shown is far less than that given. A shortened, and quite acceptable, version follows.

Example

$$u_{n+2} - 5u_{n+1} + 6u_n = 0 \qquad u_0 = 3, \ u_1 = 7$$

Solution

Try $u_n = Ax^n$. The auxiliary equation is $x^2 - 5x + 6 = 0$

$$\Rightarrow \quad (x - 2)(x - 3) = 0$$

$$\Rightarrow \quad x = 2 \ \text{or} \ 3$$

$$\Rightarrow \quad u_n = A2^n + B3^n$$

$n = 0 \qquad\qquad 3 = A + B$ \hfill (1)

$n = 1 \qquad\qquad 7 = 2A + 3B$ \hfill (2)

$(2) - 2(1) \ \Rightarrow \quad 1 = B, \ 2 = A$

$$\Rightarrow \quad u_n = 2 \times 2^n + 3^n$$

$$= 2^{n+1} + 3^n$$

It may be that the auxiliary equation cannot be solved by factorizing, in which case it will be necessary to complete the square or use the fact that

$$ax^2 + bx + c = 0 \ \Rightarrow \ x = \frac{-b \pm \sqrt{b^2 - 4ac}}{2a}$$

Example (Fibonacci)

    1, 1, 2, 3, 5, 8, 13, ... $u_0 = 1, u_1 = 1$

Solution

$$u_{n+2} = u_{n+1} + u_n$$

$$\Rightarrow \quad u_{n+2} - u_{n+1} - u_n = 0$$

Trying $u_n = Ax^n$, the auxiliary equation is

$$x^2 - x - 1 = 0$$

$$\Rightarrow \quad x = \frac{1 \pm \sqrt{1^2 - 4 \times 1 \times -1}}{2 \times 1}$$

$$\Rightarrow \quad x = \frac{1 \pm \sqrt{5}}{2}$$

General solution $\qquad u_n = A\left(\frac{1+\sqrt{5}}{2}\right)^n + B\left(\frac{1-\sqrt{5}}{2}\right)^n$

$n = 0 \quad \Rightarrow \quad 1 = A + B \hfill (1)$

$n = 1 \quad \Rightarrow \quad 1 = A\left(\frac{1+\sqrt{5}}{2}\right) + B\left(\frac{1-\sqrt{5}}{2}\right)$

$\qquad \Rightarrow \quad 2 = A\left(1+\sqrt{5}\right) + B\left(1-\sqrt{5}\right)$

subtract (1) $\Rightarrow \quad 1 = A\sqrt{5} - B\sqrt{5}$

$\qquad \Rightarrow \quad \dfrac{1}{\sqrt{5}} = A - B \hfill (2)$

$(1) + (2) \Rightarrow \quad 1 + \dfrac{1}{\sqrt{5}} = 2A$

$\qquad \Rightarrow \quad 2A = \dfrac{\sqrt{5}+1}{\sqrt{5}}$

$\qquad \Rightarrow \quad A = \left(\dfrac{\sqrt{5}+1}{2\sqrt{5}}\right)$

$(1) - (2) \Rightarrow \quad 2B = 1 - \dfrac{1}{\sqrt{5}} = \dfrac{\sqrt{5}-1}{\sqrt{5}}$

$\qquad \Rightarrow \quad B = \left(\dfrac{\sqrt{5}-1}{2\sqrt{5}}\right)$

$\qquad \Rightarrow \quad u_n = \left(\dfrac{\sqrt{5}+1}{2\sqrt{5}}\right)\left(\dfrac{1+\sqrt{5}}{2}\right)^n + \left(\dfrac{\sqrt{5}-1}{2\sqrt{5}}\right)\left(\dfrac{1-\sqrt{5}}{2}\right)^n$

$\qquad = \dfrac{1}{\sqrt{5}}\left(\dfrac{1+\sqrt{5}}{2}\right)^{n+1} - \dfrac{1}{\sqrt{5}}\left(\dfrac{1-\sqrt{5}}{2}\right)^{n+1}$

(ii) Equal real roots

Example

$$u_{n+2} - 10u_{n+1} + 25u_n = 0 \qquad u_0 = 1, u_1 = 25$$

Solution

Try $\qquad u_n = Ax^n$ which gives an auxiliary equation

$$x^2 - 10x + 25 = 0$$

$$\Rightarrow \qquad (x - 5)^2 = 0$$

$$\Rightarrow \qquad x = 5$$

Now $u_n = A \times 5^n + B \times 5^n$ does **not** work as a general solution

$n = 0 \qquad \Rightarrow \qquad u_0 = A + B = 1$ \hfill (1)

$n = 1 \qquad \Rightarrow \qquad u_1 = 5A + 5B = 25$ \hfill (2)

$(2) - 5(1) \qquad \Rightarrow \qquad 0 = 20$?

These equations (1) and (2) are inconsistent and so there is no finite solution.

Try $\qquad u_n = A \times 5^n + Bn \times 5^n$

$n = 0 \qquad \Rightarrow \qquad u_0 = A + 0 = 1$ \hfill (1)

$n = 1 \qquad \Rightarrow \qquad u_1 = 5A + 5B = 25$ \hfill (2)

$(2) + 5 \qquad \Rightarrow \qquad A + B = 5$

$A = 1 \qquad \Rightarrow \qquad B = 4$

$$u_n = 5^n + 4n \times 5^n$$

In general, $\qquad au_{n+2} + bu_{n+1} + cu_n = 0$

$$\Rightarrow \qquad u_n = AX_1^n + BX_2^n \text{ where } X_1, X_2$$

$$\text{are different real roots of } ax^2 + bx + c = 0$$

$$\text{and } u_n = AX_1^n + BnX_1^n \text{ when } X_1 = X_2.$$

Exercise 7.1   Solve the following second order recurrence relations.

(a) $u_{n+2} - 2u_{n+1} - 8u_n = 0$ \qquad\qquad $u_0 = 8, u_1 = 14$

(b) $2u_{n+2} - 5u_{n+1} - 3u_n = 0$ \qquad\qquad $u_0 = 11, u_1 = 5$

(c) $u_{n+2} = 6u_{n+1} - 9u_n$ \qquad\qquad $u_0 = 1, u_1 = 9$

(d) $4u_{n+2} + u_n = 4u_{n+1}$ \qquad\qquad $u_0 = 3, u_1 = 1$

Exercise 7.2   Hardwick brain cells have a life span of two seconds. At the end of two seconds each cell splits into 6 new cells. Let c_n be the number of cells after n seconds and write an equation giving c_n in terms of c_{n-1} and c_{n-2}.

Given that $c_0 = 8$ and $c_1 = 9$, solve your equation to find an expression for c_n. How many cells are there after 15 seconds?

Exercise 7.3   Solve $u_{n+2} = 4u_n$ $\qquad\qquad$ $u_0 = 3, u_1 = 2$

Exercise 7.4   Solve $u_{n+2} - 3u_n = 0$ $\qquad\qquad$ $u_0 = 1, u_1 = 9\sqrt{3}$

Remember $u_{n+2} - 10u_{n+1} + 25u_n = 0$ is the same as $u_n - 10u_{n-1} + 25u_{n-2} = 0$.

12.8 SECOND ORDER HOMOGENEOUS
Example

$$u_{n+2} - u_{n+1} - 12u_n = 60n - 17 \qquad u_0 = 6, u_1 = 2$$

Solution

Consider: $u_{n+2} - u_{n+1} - 12u_n = 0$ and try $u_n = Ax^n$

auxiliary equation:

$$x^2 - x - 12 = 0$$
$$\Rightarrow \quad (x - 4)(x + 3) = 0$$
$$\Rightarrow \quad x = 4 \text{ or } -3$$
$$\Rightarrow \quad u_n = A \times 4^n + B(-3)^n \text{ is the complementary function.}$$

Try a particular solution $Cn + D$ as for first order equations.

General solution:

$$u_n = A \times 4^n + B(-3)^n + Cn + D$$
$$u_{n+2} - u_{n+1} - 12u_n = 60n - 17$$

$n = 0$ $\quad \Rightarrow \quad$ $u_2 - u_1 - 12u_0 = 60 \times 0 - 17$
$\qquad\qquad \Rightarrow \quad$ $u_2 - 2 - 72 = -17$
$\qquad\qquad \Rightarrow \quad$ $u_2 = 57$

$n = 1$ $\quad \Rightarrow \quad$ $u_3 - u_2 - 12u_1 = 60 \times 1 - 17$
$\qquad\qquad \Rightarrow \quad$ $u_3 - 57 - 24 = 43$
$\qquad\qquad \Rightarrow \quad$ $u_3 = 124$

$$n = 0 \quad \Rightarrow \quad 6 = A + B + D \quad\quad (1)$$

$$n = 1 \quad \Rightarrow \quad 2 = 4A - 3B + C + D \quad\quad (2)$$

$$n = 2 \quad \Rightarrow \quad 57 = 16A + 9B + 2C + D \quad\quad (3)$$

$$n = 3 \quad \Rightarrow \quad 124 = 64A - 27B + 3C + D \quad\quad (4)$$

$$(2)-(1) \quad\quad -4 = 3A - 4B + C \quad\quad (5)$$

$$(3)-(2) \quad\quad 55 = 12A + 12B + C \quad\quad (6)$$

$$(4)-(3) \quad\quad 67 = 48A - 36B + C \quad\quad (7)$$

$$(6)-(5) \quad\quad 59 = 9A + 16B \quad\quad (8)$$

$$(7)-(6) \quad\quad 12 = 36A - 48B \quad\quad (9)$$

$$(8)+(9)\div 3 \Rightarrow \quad 63 = 21A$$

$$\Rightarrow \quad A = 3, \quad B = 2, \quad C = -5, \quad D = 1$$

$$\Rightarrow \quad u_n = 3 \times 4^n + 2(-3)^n - 5n + 1$$

12.9 SUMMARY
First order linear

homogeneous:

$$u_{n+1} = au_n \quad \Rightarrow \quad u_n = a^n u_0$$

inhomogeneous $(a \neq 1)$:

$$u_{n+1} = au_n + b \quad \Rightarrow \quad u_n = A(a)^n + B$$

$$= au_n + bn + c \quad \Rightarrow \quad u_n = A(a)^n + Bn + C$$

$$= au_n + bn^2 + cn + d \quad \Rightarrow \quad u_n = A(a)^n + Bn^2 + Cn + D$$

$$= au_n + b\lambda^n \quad \Rightarrow \quad u_n = A(a)^n + B\lambda^n$$

$$u_{n+1} = u_n + f(n) \quad \Rightarrow \quad u_n = g(n)$$

where g has degree one higher than that of f.

Second order linear

homogeneous:

$$au_{n+2} + bu_{n+1} + cu_n = 0 \quad \Rightarrow \quad u_n = AX_1^n + BX_2^n$$

where X_1, X_2 are distinct real roots of $ax^2 + bx + c = 0$

or $\quad \Rightarrow \quad u_n = AX_1^n + BnX_1^n$

where X_1 is the only real root of $ax^2 + bx + c = 0$.

inhomogeneous: as first order linear homogeneous $(a \neq 1)$.

Exercise 9.1   Solve $u_{n+2} - 5u_{n+1} + 6u_n = 2$ $u_0 = 0, u_1 = 1$

Exercise 9.2   Solve $u_{n+2} - 4u_n = 9$ $u_0 = 2, u_1 = 5$

12.10 MISCELLANEOUS EXERCISES
Solve the following:

Exercise 10.1 $u_{n+1} - 7u_n = 0$ $u_0 = 3$

Exercise 10.2 $u_n - u_{n-1} = 5$ $u_0 = 0$

Exercise 10.3 $u_{n+1} - 2u_n = 3$ $u_0 = -1$

Exercise 10.4 $u_{n+1} - 0.2u_n = 0.5^n$ $u_0 = 5$

Exercise 10.5 $5u_n - 20u_{n-1} = 3$ $u_0 = 6$

Exercise 10.6 $u_n - 13u_{n-1} + 36u_n = 0$ $u_0 = -1, u_1 = 1$

Exercise 10.7 $9u_{n+2} + 6u_{n+1} + u_n = 0$ $u_0 = 5, u_1 = -2$

Exercise 10.8   Explain why the general solution of $u_{n+2} - 4u_{n+1} + 4u_n = n$ is of the form
$u_n = (A + Bn)2^n + Cn + D$. Write down similar expressions for u_{n+1} and u_{n+2} and substitute these into the original equation. Hence write down the values of C and D.

Exercise 10.9   Solve $u_n - 5u_{n-1} + 6u_{n-2} = 14$ by the method of Exercise 10. 8, leaving two constants in your answer.

Exercise 10.10 Solve $u_{n+1} = 2u_n + 5(3)^n$, $u_0 = 1$

Exercise 10.11 θ particles split into 3 new θ particles every second. ϕ particles divide every second to produce one θ particles and 2 new ϕ particles. If θ_n is the number of θ particles after n seconds and ϕ_n is defined similarly, complete the two recurrence relations

        (a) $\phi_n =$

and   (b) $\theta_n =$

Given that $\phi_0 = 1$, solve (a) to give a formula for ϕ_n.

Given also that $\theta_0 = 1$ and using your previous answer, obtain an expression for θ_n.

What happens to θ_n as $n \to \infty$?

Exercise 10.12 What is the maximum number of regions that can be created with n lines? With three lines the number is 7. When a fourth line is added it will cut the three lines already drawn. This is shown in Fig. 12.3, where the fourth line will go through the points marked A, B and C. Since this line cuts three lines it must go through four regions which it divides, so there will be four more than previously ($r_4=11$). Write down a recurrence relation connecting r_{n+1} and r_n. Using the fact that $r_0=1$, solve your equation. What is the maximum number of regions that can be created with 100 lines?

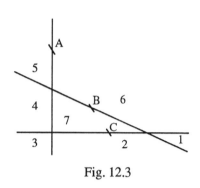

Fig. 12.3

Exercise 10.13 You have borrowed £30 000 from a building society. Interest is charged at 12% per annum on the amount owing at the beginning of the year and you are required to make repayments of £M per month. Write down, but do not evaluate, an expression for the amount owing after one year.

Show that $P_n - \lambda P_{n-1} = -12M$ where P_n is the amount owing after n years and give the value of λ. If the loan is to be paid off in 25 years, solve the recurrence relation to find a general expression for P_n and also the value of M. (Take $1.12^{25} = 17$.)

Exercise 10.14 Robert was given a bottle of fruit drink. Each day he drank one quarter of its contents and, so it did not appear that he was drinking it so quickly, he added just enough water to raise the level of the liquid half-way to the top.

If ω_n = the amount of actual fruit drink in the bottle after n days, write down and solve a recurrence relation for ω_n, given that $\omega_0 = 1$. If t_n = the total amount of liquid in the bottle after n days, show that $t_{n+1} - \dfrac{3}{8}t_n = \dfrac{1}{2}$. Solve this recurrence relation to give t_n. What can you say about ω_n and t_n as $n \to \infty$?

Exercise 10.15 Water is used as a coolant for a certain industrial process. The tank holding the water has a capacity of 16 units and initially contains all fresh water. Each day one unit of water is lost by evaporation. In order to prevent the build-up of a dangerously high concentration of impurities in the water, at the end of each day a further three units of water are removed from the tank. Then four units of fresh water are added to restore the total volume of coolant to 16 units. The concentration of impurities in fresh water is c kg per unit of water. If u_n denotes the number of kg of impurities in the tank at the end of the nth day, after the water level has been brought back to normal, show that $u_{n+1} - \dfrac{4}{5}u_n = 4c$. What can you say about u_n as $n \to \infty$?

Exercise 10.16 The demand for places at a certain school depends on the percentage fee increase the previous year. The increase each year depends on the demand for places at the start of that year and has the form

$$P_t = 4 + \frac{D_t}{50}$$

where P_t is the percentage increase in the t th year and D_t is the demand for places that year.

It is found over a period of time that $D_t = 300 - 5P_{t-1}$.

Form a first order recurrence equation for D_t. Solve your equation, leaving one constant undetermined. Find the long term prospects for demand and use this to find the likely fee increase then.

12.11 NOTES
The word **recurrence** comes from the Latin word *currere* which means *to run*.

Geometric progressions

The terms of a geometric progression (GP) $a, ar, ar^2, ...,$

can be described by $u_{n+1} = ru_n$ $u_1 = a$

$\Rightarrow$ $u_n = ar^{n-1}$

The sum of the first n terms is given by

$$S_{n+1} = S_n + ar^n \qquad S_1 = a$$

$\Rightarrow \quad S_n = A + Br^n$

$S_1 = a = A + Br$ .... (1)

$S_2 = a + ar = A + Br^2$ .... (2)

subtract $ar = Br^2 - Br$

$a = Br - B = B(r-1)$

$\Rightarrow \quad B = \dfrac{a}{r-1}$

(1) $\Rightarrow \quad a = A + \dfrac{ar}{r-1}$

$\Rightarrow \quad \dfrac{ar - a}{r-1} = A + \dfrac{ar}{r-1}$

$\Rightarrow \quad A = \dfrac{-a}{r-1}$

$\Rightarrow \quad S_n = -\dfrac{a}{r-1} + \dfrac{ar^n}{r-1}$

$$= \dfrac{a(r^n - 1)}{r-1}$$

Arithmetic progressions

The terms of an arithmetic progression (AP) $a, a+d, a+2d, \ldots$

can be described by

$$u_{n+1} = u_n + d \qquad u_1 = a$$

$$\Rightarrow \quad u_n = A + Bn$$

$$u_1 = a = A + B$$

$$u_2 = a + d = A + 2B$$

$$\Rightarrow \quad B = d, \ A = a - d$$

$$\Rightarrow \quad u_n = a - d + dn$$

$$= a + (n-1)d$$

The sum of the first n terms is given by

$$S_{n+1} = S_n + a - d + dn$$

$$\Rightarrow \quad S_n = A + Bn + Cn^2$$

$$n = 1 \qquad a = A + B + C \tag{1}$$

$$n = 2 \qquad 2a + d = A + 2B + 4C \tag{2}$$

$$n = 3 \qquad 3a + 3d = A + 3B + 9C \tag{3}$$

$$(2)-(1) \Rightarrow \quad a + d = B + 3C \tag{4}$$

$$(3)-(2) \Rightarrow \quad a + 2d = B + 5C \tag{5}$$

$$(5)-(4) \Rightarrow \quad d = 2C$$

$$\Rightarrow \quad C = \frac{d}{2}, \ B = a - \frac{d}{2}, \ A = 0$$

$$S_n = \left(a - \frac{d}{2}\right)n + \frac{d}{2}n^2$$

$$= an - \frac{d}{2}n + \frac{dn^2}{2}$$

$$= \frac{n}{2}(2a - d + dn)$$

$$= \frac{n}{2}\left[2a + (n-1)d\right]$$

Second order linear – auxiliary equation has no real roots
A knowledge of complex numbers is required for this type of equation.

Example

$$u_{n+2} - 2u_{n+1} + 2u_n = 3 \qquad u_0 = 4, \ u_1 = 6$$

$$\Rightarrow \quad u_2 = 7$$

Consider $\quad u_{n+2} - 2u_{n+1} + 2u_n = 0$

Try $u_n = A\lambda^n$ so the auxiliary equation is $x^2 - 2x + 2 = 0$

$$x = \frac{2 \pm \sqrt{(-2)^2 - 4 \times 1 \times 2}}{2 \times 1}$$

$$= \frac{2 \pm \sqrt{-4}}{2}$$

$$= 1 \pm \sqrt{-1}$$

$$X_1 = 1 + i, \ X_2 = 1 - i$$

$$\Rightarrow \quad |X_1| = |X_2| = \sqrt{2}$$

$$\arg X_1 = \frac{\pi}{4} \qquad \arg X_2 = -\frac{\pi}{4}$$

$$\Rightarrow \quad X_1 = \sqrt{2} \, e^{\frac{i\pi}{4}} \qquad\qquad X_2 = \sqrt{2} \, e^{-\frac{i\pi}{4}}$$

$$u_n = A\left(\sqrt{2} \, e^{\frac{i\pi}{4}}\right)^n + B\left(\sqrt{2} \, e^{-\frac{i\pi}{4}}\right)^n$$

$$= A2^{\frac{n}{2}} e^{\frac{in\pi}{4}} + B e^{\frac{\pi}{2}} e^{-\frac{in\pi}{4}}$$

$$= 2^{\frac{n}{2}}\left(A e^{\frac{in\pi}{4}} + B e^{-\frac{in\pi}{4}} \right)$$

$$= 2^{\frac{n}{2}}\left\{ P\cos\left(\frac{n\pi}{4}\right) + Q\sin\left(\frac{n\pi}{4}\right) \right\}$$

General solution $u_n = 2^{\frac{n}{2}} \left\{ P \cos\left(\frac{n\pi}{4}\right) + Q \sin\left(\frac{n\pi}{4}\right) \right\} + R$

$n = 0$ $4 = P + R$ (1)

$n = 1$ $6 = \sqrt{2}\left(P\frac{1}{\sqrt{2}} + Q\frac{1}{\sqrt{2}} \right) + R$

$\Rightarrow$ $6 = P + Q + R$ (2)

$n = 2$ $7 = 2(P \times 0 + Q \times 1) + R$

$\Rightarrow$ $7 = 2Q + R$

                                                                            (3)

$(2) - (1) \Rightarrow \quad 2 = Q$

$(3) \quad \Rightarrow \quad R = 3$

$(1) \quad \Rightarrow \quad P = 1$

$\Rightarrow \quad u_n = 2^{\frac{n}{2}}\left(\cos\frac{n\pi}{4} + 2\sin\frac{n\pi}{4} \right) + 3$

R could also be found by substituting into the recurrence equation,

$$R - 2R + 2R = 3$$

$\Rightarrow \quad R = 3$

This method of solution is quicker as there is no need to find u_2.

The general solution may also be written as

$$u_n = 3 + \sqrt{5} \times 2^{\frac{n}{2}} \cos\left(\frac{n\pi}{4} - 1.1\right)$$

$n \rightarrow \infty \Rightarrow 2^{\frac{n}{2}} \rightarrow \infty$, so the values of u_n oscillate about 3 with increasing amplitude:

    4, 6, 7, 5, −1, −9, −13, −5, 19, 51, 67, 35, 61, ...

13

Simulation

13.1 INTRODUCTION

To **simulate** is to **imitate** the conditions of a situation.

For example, if a shop manager employing 10 cashiers wanted to cut costs by reducing the number employed on checkouts to 8, a simulation could be used to examine what queue lengths might develop. If long queues were caused by trying it in the shop, this would not be good for customer relations!

Radioactive decay can be modelled using drawing pins with each one representing an atom of the material. They are tipped onto a flat surface and each one that is seen to be point down is considered to have decayed and, therefore, removed. Those still active are tipped out again and at each stage the number of such 'atoms' is noted.

13.2 RANDOM NUMBERS

Most of the simulations we will deal with are based on the use of **random numbers**, so what are 'random numbers'? Most calculators will produce random numbers from 0.000 to 0.999 by using the RAN key. However, these thousand numbers and those generated by computers are not truly random as they result from the execution of functions programmed into the machines. Long strings of these numbers do not have any discernable pattern so prediction of values is not possible, thus making them seem random.

Statisticians have pages of random numbers in most text books and sets of statistical tables. They are useful in a variety of situations.

You can generate truly random numbers for yourself very simply in a variety of ways. For example, make a ten-sided spinner with its edges labelled 0 to 9 and spin it three times to produce a number from 000 to 999.

You can toss coins to produce random numbers. Recording a head (H) as 1 and a tail (T) as 0, the sequence HTTH would, using the binary system, give the number

$$2^4 \quad 2^2 \quad 2 \quad 2^0$$

$$1 \quad\ \ 0 \quad\ \ 0 \quad 1 \qquad = 9$$

Using 4 coins you can generate sixteen random numbers (0-15).

13.3 USE OF RANDOM NUMBERS
Example

Records over a long period show that a darts player has a probability of winning a match of 0.75 if he won his previous one and a probability of 0.45 of losing if he lost his previous one.

(a) Using the random numbers below, simulate his results for a period of 25 matches, given that the match prior to the simulation was a loss.

$$56 \quad 00 \quad 36 \quad 35 \quad 97 \quad 12 \quad 75 \quad 67 \quad 90$$
$$77 \quad 78 \quad 28 \quad 71 \quad 51 \quad 93 \quad 47 \quad 30 \quad 73$$
$$34 \quad 97 \quad 79 \quad 88 \quad 95 \quad 39 \quad 49$$

(b) From your simulation, estimate the overall proportion of matches won.

(c) Use your results to produce a table of frequencies of occurrences of runs of one, two, three, etc. wins. For example, ... LWWWWL... represents a run of four wins.

Represent your table of frequencies on a bar chart.

(d) The darts player wins the opening match of a tournament. Estimate the probability that he will also win in the next three rounds.

Solution

Following a win (W), the numbers 00 to 74 mean that the next result is a win
and the numbers 75 to 99 mean that the next result is a loss.

Following a loss (L), the numbers   0 to 44 mean that next result is a loss
and the numbers 45 to 99 mean that the next result is a win.

The random numbers can be taken from the table in any appropriate way. For this example, assume that the random numbers from (a) above are read across (from left to right) each row in turn.

(a) prior result L $\Rightarrow$ W   W   W   W   L   L   W   W   L

W   L   L   W   W   L   W   W   W

W   L   W   L   W   W   W

(b)   proportion won $= \dfrac{17}{25} = 0.68$

(c)

run length	frequency
1	2
2	2
3	1
4	2

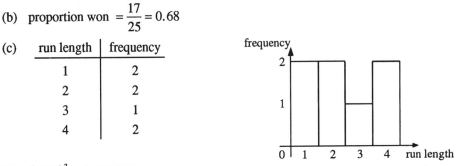

(d) $(0.75)^3 = 0.421875$

$= 0.422$ (to 3 significant figures)

Example

At a funfair punters can have 5 shots on a rifle range for £1. On average 20% of shots miss the target.

(a)   Use the random digits below to simulate a hit (H) or a miss (M) for each shot, for fifty shots.

76	33	84	92	71	68	39	19	39	25
84	35	93	65	79	02	28	56	83	24
97	08	59	27	07	30	31	02	83	22
88	00	06	08	68	77	26	98	88	86
82	72	42	32	48	20	01	58	77	04

(b)   Group your results in sets of 5, and thus count how many hits are recorded for each person.

(c)   Using your results from (b), estimate
(i) the probability that a prize will be won if five hits are needed to win;
(ii) the mean number of hits per person.

(d)   Calculate the theoretical probability of a punter scoring 5 hits.

(e)   Explain the difference between your answers to (c)(i) and (d).

Solution

(a) and (b)   Using 0-79 for a hit (H) and 80-99 for a miss a (M), the following table will be generated.

H	H	M	M	H	3 hits	H	H	H	H	H	5 hits
M	H	M	H	H	3 hits	H	H	H	M	H	4 hits
M	H	H	H	H	4 hits	H	H	H	M	H	4 hits
M	H	H	H	H	4 hits	H	H	M	M	M	2 hits
M	H	H	H	H	4 hits	H	H	H	H	H	5 hits

(c) (i) $P(\text{prize won}) = \dfrac{2}{10} = 0.2$

   (ii) $(3+3+4+4+4+5+4+4+2+5) \div 10 = 3.8$

(d) $0.8^5 = 0.32768$

(e) With such a small sample the answers should be expected to be different. Only with large samples would you expect approximate equality.

There is no need for all random numbers to be assigned.

Example

When Sarah and Paul play darts Sarah wins one third of the games and Paul wins the rest. How would you use random numbers to simulate the situation?

Solution

Either   use the numbers 01 to 33 to represent Sarah winning and use the numbers 34 to 99 to represent Paul winning. 00 is discarded;

or   consider the digits one at a time. Use the numbers 1 to 3 to represent Sarah winning and use 4 to 9 to represent Paul winning. 0 is discarded.

Exercise 3.1   Personnel Incorporated is an agency which finds people for top jobs. Of those who apply for posts, 30% are rejected based on their letters of application. Of the remainder, 30% fail the first interview and, of those who go on to the aptitude test, 25% do not go forward to the final interview.

(a) Using the random digits below, simulate the application success or otherwise of ten candidates. (Work along rows.)

$$41 \quad 95 \quad 96 \quad 86 \quad 70 \quad 45 \quad 27 \quad 48 \quad 38 \quad 80$$
$$07 \quad 09 \quad 25 \quad 23 \quad 92 \quad 24 \quad 62 \quad 71 \quad 06 \quad 07$$
$$06 \quad 55 \quad 84 \quad 53 \quad 44 \quad 67 \quad 33 \quad 84 \quad 53 \quad 20$$

(b) How many applicants were given a final interview?

(c) Calculate the theoretical probability of a candidate getting a final interview.

(d) Theoretically, how many of the ten would you expect to get a final interview?

Exercise 3.2   (a) A machine produces glass plate; 65% of its output has no flaws. In a production run of 100 sheets simulate, using the digits given, whether each is perfect (P) or flawed (F).

$$
\begin{array}{cccccccccc}
87 & 89 & 26 & 74 & 07 & 19 & 56 & 67 & 07 & 92 \\
68 & 21 & 27 & 69 & 82 & 78 & 36 & 30 & 32 & 50 \\
07 & 20 & 38 & 96 & 22 & 62 & 74 & 20 & 56 & 09 \\
49 & 25 & 89 & 11 & 68 & 42 & 53 & 43 & 75 & 29 \\
51 & 46 & 99 & 69 & 19 & 92 & 58 & 02 & 59 & 13 \\
31 & 47 & 18 & 04 & 39 & 23 & 27 & 20 & 05 & 73 \\
64 & 24 & 13 & 74 & 67 & 63 & 06 & 18 & 85 & 31 \\
59 & 05 & 04 & 14 & 30 & 55 & 73 & 52 & 62 & 20 \\
93 & 91 & 75 & 77 & 08 & 76 & 38 & 29 & 74 & 93 \\
94 & 45 & 16 & 47 & 56 & 33 & 67 & 16 & 47 & 80 \\
\end{array}
$$

(b) Count the lengths of the runs of perfect glass plates and note them (... F P P P F ... is a run of three). Show your results in a histogram. (FF is a run of 0 as no Ps appear between 2 Fs.)

(c) Explain why the probability of a run of five perfect items is

$$0.65^5 \times 0.35$$

(d) Calculate probabilities of runs of length (i) 0 (ii) 1 (iii) 2 (iv) 5 (v) 13.

(e) Compare these results with your histogram.

Exercise 3.3   Two competitors in a 'Master Brain' competition have to solve puzzles in under a minute each. The probability that Jo will succeed is 0.9 while the probability that Sam will do so is 0.85.

(a) Using the following random digits simulate their performances over a series of 30 puzzles.

Use for Jo: 47 57 36 73 97 89 54 51 48 26
                 33 80 18 77 39 92 93 01 36 96
                 01 04 26 33 77 62 95 61 39 77

Use for Sam: 03 71 87 11 42 22 96 37 98 29
                 23 17 92 86 84 95 44 00 32 98
                 27 94 15 78 23 61 07 03 67 27

(b) How many times in the 30 goes do (i) neither (ii) just one (iii) both, succeed?

(c) Calculate the theoretical probabilities of (i) 0 (ii) 1 (iii) 2, successes.

(d) If they play three times in each of ten rounds, how many times, according to your simulation, would they both succeed every time?

(e)   Find the theoretical probability that they will each have three successes in a round.

(f)   Compare your answers to (d) and (e).

Exercise 3.4   A and B play darts against each other in a competition to hit the 'bulls-eye'.
The probability that A succeeds with a shot is 0.6 and that B does is 0.5.

To win a **game** a player must   (a) succeed at least 3 times, and
                                      (b) be at least two ahead of the opponent.

To win a **set** a player must   (a) win at least 5 games, and
                                      (b) win at least two more games than the opponent.

The first to win three sets has won the match.

Using, row by row, the random numbers below, simulate a game between A and B given
that A starts.

Continue the simulation of a set if the players take turns to start the games. Show who
wins each game in the set, and who wins the set.

```
23  40  29  82  73  04  79  47  62  79
84  35  93  65  79  13  97  92  79  99
33  11  71  21  94  37  76  29  02  51
77  26  98  00  56  82  72  42  82  48
22  18  27  94  15  78  23  61  07  03
67  27  18  74  33  47  63  78  16  97
76  25  96  24  76  01  04  26  33  77
```

Exercise 3.5   Bars of chocolate are sold in packs of four. During a special promotion one
eighth of all individual bars carry a coupon entitling the purchaser to a 10p refund on the cost.

(a)   Use the random numbers provided here to simulate a bar having or not having a coupon,
      for twenty five packs.

```
61187  77044  65793  30946  54843  00567
83826  72024  25558  61874  27648  01008
89826  99021  47977  22380  00085  80109
67513  99566  11005  56442  71968  37977
51330  96175  75898  43465  36411  42832
43955  57999  84124  85368  14308  24423
```

(b)   How many packs contain (i) 0 (ii) 1 (iii) 2 (iv) 3 (iv) 4 (vi) 5 coupons?

(c)   What is the mean number of coupons per pack?

(d)   Calculate the theoretical probability of no coupons in a pack.

(e)   How many of the twenty five packs would you expect to contain no coupons?

13.4 NOTES

Random number may be used to model any situation for which the theoretical probability distribution is known and the cumulative distribution function can be calculated or is given in tables.

Example

At a small supermarket 40% of evening shoppers are male and 60% are female. Five of the checkouts are kept in operation throughout this period and X is 'the number of males in a sample of the five most recently served customers, taking the last at each till'.

By calculation or from cumulative binomial tables with $p = 0.4$, giving answers correct to 3 decimal places,

$$P(X \le 0) = 0.078 \qquad P(X \le 3) = 0.913$$
$$P(X \le 1) = 0.337 \qquad P(X \le 4) = 0.990$$
$$P(X \le 2) = 0.682 \qquad P(X \le 5) = 1.000$$

Now the random number 0.721 given on a calculator produces the value $X = 3$.

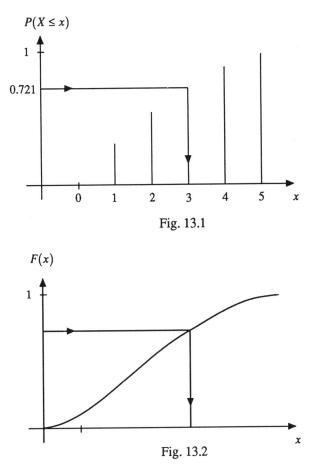

This process can form the basis of computer programs to simulate such situations.

Fig. 13.1

Cumulative tables of continuous distributions can be used in the same way.

Fig. 13.2

14

Iterative processes

14.1 INTRODUCTION

How could you set about solving the equation $x^5 + x - 3 = 0$?

This does not look very different from $x^2 + x - 3 = 0$, which does not pose a problem as you have a formula available that will give the answers to quadratic equations. But what about equations containing higher powers? Is it possible to solve cubics, quartics, quintics, etc., by radicals? That is, can the roots of such equations be found by a formula or set of formulae involving only polynomials? **Cardano's formula** (see section 14.10) provides solutions to cubic equations and his student Ferrari went one step further to solve quartics. However, no similar results are available for quintics or higher powers. It is not that they are yet to be discovered but rather that they do not exist at all, as is known from the work of the remarkable French algebraist Galois.

So what is needed is some **method** for solving quintics and higher order equations, as no formula is available. Indeed, there is no reason why such methods should not be applied to cubics and quartics since Cardano's formula is certainly not trivial to use and Ferrari's is one level above this in difficulty.

14.2 INVESTIGATION

(a) Starting with $x_1 = 1$, use the following rules to produce sequences as far as x_{10} where possible. (x_3 is given in each case as a check.)

(i) $x_n = \sqrt[5]{3 - x_{n-1}}$ (1.1311)

(ii) $x_n = \sqrt{\dfrac{3 - x_{n-1}}{x_{n-1}^3}}$ (0.7488)

(iii) $x_n = \sqrt{\sqrt{\dfrac{3}{x_{n-1}} - 1}} = \sqrt[4]{\dfrac{3}{x_{n-1}} - 1}$ (1.1108)

(iv) $x_n = 3 - x_{n-1}^5$ (−29)

(b)   Which of your values of x_{10} are solutions of

$$x^5 + x - 3 = 0$$

to a reasonable degree of accuracy?

(c)   Why are they solutions to this equation?

(d)   Are the sequences that produce the solution equally quick at moving towards it or are some methods better than others?

14.3 CONVERGENT SEQUENCES

An **iterative process** is one which is repeated. In the sequences generated in section 14.2, you have the results of four iterations, that is, repetitions of processes described by the four formulae.

In two of the cases, (a)(i) and (a)(iii), the sequences of numbers **converge**, or become closer to, a finite limit. In a(iv) the sequence **diverges** (the opposite of converged), and produces answers so large that calculators soon cannot cope. (a)(ii) diverges for a while, as it shows no sign of converging to any finite number, and then self-destructs as x_8 is not real and so it terminates at x_7.

The two that converge do so to the same number. By x_8 they are equal to three decimal places and by x_{10} they are the same, correct to four decimal places. They are both moving towards the solution of $x^5 + x - 3 = 0$. But why?

Suppose the sequence produced by (a)(i) is leading towards some finite number x. This is the value which, when substituted into the right-hand side, produces itself as the answer.

So $x = \sqrt[5]{3 - x}$

$\Rightarrow$ $x^5 = 3 - x$

$\Rightarrow$ $x^5 - x + 3 = 0$

Since the value produced works in the first line, it also works in the third which is just a rearrangement of it.

What of (a)(ii), (a)(iii) and (a)(iv)?

If they converged they would also produce solutions to the equation $x^5 + x - 3 = 0$ since (a)(ii) $x = \sqrt{\dfrac{3-x}{x^3}}$ $\qquad$ (a)(iii) $x = \sqrt[4]{\dfrac{3}{x} - 1}$ $\qquad$ (a)(iv) $x = 3 - x^5$

$\Rightarrow x^2 = \dfrac{3-x}{x^3}$ $\qquad$ $\Rightarrow x^4 = \dfrac{3}{x} - 1$ $\qquad$ $\Rightarrow x^5 + x - 3 = 0$

$\Rightarrow x^5 = 3 - x$ $\qquad$ $\Rightarrow x^5 = 3 - x$

$\Rightarrow x^5 + x - 3 = 0$ $\qquad$ $\Rightarrow x^5 + x - 3 = 0$

So each of the iterative formulae is a rearrangement of the original equation. Some produce solutions to it but others do not and those that do succeed do not approach the answer at the same rate.

To solve $x^5 + x - 3 = 0$ you could rearrange as follows:

$$x^5 = 3 - x$$

$$x = \frac{3-x}{x^4}$$

and then produce a sequence using the iterative formula $x_n = \dfrac{3 - x_{n-1}}{x_{n-1}^4}$. However, you have no guarantee that the sequence produced will actually converge. If it does not, then try another.

Note $x_n = \sqrt{5x_{n-1}}$ has the same meaning as $x_{n+1} = \sqrt{5x_n}$, as each says that a term in the sequence is obtained by multiplying the previous one by 5 and taking the square root of the result.

Exercise 3.1 Use the iterative formula $x_n = \sqrt{8x_{n-1}}$ to produce a sequence starting with $x_1 = 1$ up as far as x_{10}. What equation is being solved by this process?

Exercise 3.2 Use the iterative formula $x_n = x_{n-1}(2 - 27x_{n-1})$ to produce a sequence starting with $x_1 = 0.04$ and stop when there is no change in the sixth decimal place. What number has just been found?

Exercise 3.3 Show that using the iterative formula $x_n = \dfrac{5}{2x_{n-1}^2} + 2$ is an attempt to solve the equation $2x^3 - 4x^2 - 5 = 0$.

Use the formula to produce a sequence of numbers starting with $x_1 = 2$ and if the sequence converges, give a solution to the equation, correct to 4 decimal places.

Exercise 3.4 Use the iterative formula $x_n = \dfrac{1}{2}\left(x_{n-1} + \dfrac{2}{x_{n-1}} \right)$ to find the square root of 2, correct to 3 decimal places and verify that the formula is a rearrangement of $x^2 = 2$.

Exercise 3.5

(a) Show that the equation $x^4 - 3x^3 - 10 = 0$ may be rearranged to give $x = \dfrac{10}{x^3} + 3$.

Hence write down an iterative formula that may solve the original equation.

(i)   Produce a sequence starting with $x_1 = 1$ and stop when successive answers are equal correct to three decimal places. Write down the answer correct to three decimal places.

(ii) Produce sequences starting with $x_1 = 100$ and $x_1 = -2$. Do these converge to the same value?

(b) Show that the equation $x^4 - 3x^3 - 10 = 0$ may be rearranged to produce the iterative

formula $x_n = \sqrt[3]{\dfrac{x_{n-1}^4 - 10}{3}}$. Use this formula to produce a solution to the equation correct

to two decimal places.

(c) What do you deduce from your results obtained in (a)(i) and (b) above?

Exercise 3.6   Solve the equation $x^3 - 2x^2 - 1 = 0$ by rearranging to produce an iteration formula. Give your answer correct to 3 decimal places.

Exercise 3.7   The cylindrical tin in Fig. 14.1 has a volume of 260 cm³. If its height is 5 cm more than its radius, show that

$$r^3 + 5r^2 - \frac{260}{\pi} = 0 \quad [\text{volume} = \pi r^2 h]$$

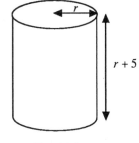

Use the iteration formula $r_n = \sqrt{\dfrac{52}{\pi} - \dfrac{r_{n-1}^3}{5}}$ with $r_1 = 3$,

to produce r_2, r_3 and r_4.

Fig. 14.1

Use $\dfrac{r_3 + r_4}{2}$ as an approximation to the solution and apply the formula until you can give

the answer correct to three decimal places. Use the idea $\dfrac{r_n + r_{n-1}}{2}$ to shorten the working

if you wish.

Exercise 3.8   Solve the equation $2^x = x^3 - 7$ by using the iteration formula

$$x_n = \sqrt[3]{2^{x_{n-1}} + 7}$$

giving your answer correct to 2 decimal places.

14.4 GRAPHICAL REPRESENTATION

Some of the iterative attempts to solve $x^5 + x - 3 = 0$ were successful and some were not. You will now investigate the values produced in the sequences by the four different formulae suggested in section 14.2.

The equation $x^5 + x - 3 = 0$ can be rearranged to give $x = \sqrt[5]{3 - x}$.

This equation could be solved graphically by finding the x-coordinate of the point of intersection of $y = \sqrt[5]{3 - x}$ and $y = x$.

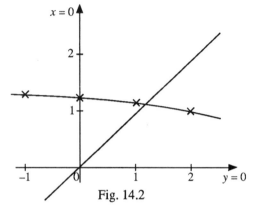

x	−1	0	1	2
$y = \sqrt[5]{3 - x}$	1.32	1.25	1.15	1

The iterative formula

$$x_n = \sqrt[5]{3 - x_{n-1}}$$

Fig. 14.2

works by taking x_1 and then evaluating $x_2 = \sqrt[5]{3 - x_1}$ which is a point, A, on the curve with coordinates (x_1, x_2). The x_1 is now discarded and replaced by x_2 giving the point B on the straight line. Now x_2 is used to calculate $x_3 = \sqrt[5]{3 - x_2}$ and the point (x_2, x_3), denoted by C, is on the curve. The x_2 is replaced by x_3 to give the point D with coordinates (x_3, x_3).

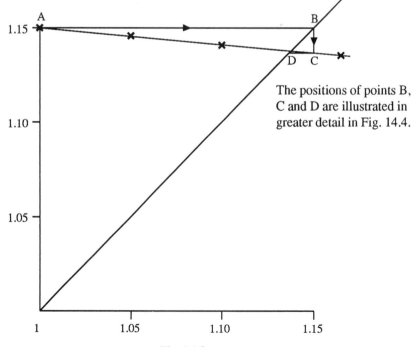

The positions of points B, C and D are illustrated in greater detail in Fig. 14.4.

Fig. 14.3

So you move horizontally from the curve to the line and vertically from the line to the curve, spiralling in towards the point of intersection you seek.

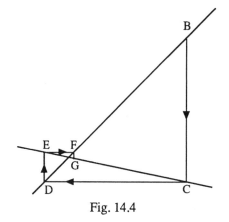

Fig. 14.4

This pattern is followed as long as the curve has a gradient, g, satisfying $-1 < g < 0$ at the point of intersection.

The closer the gradient is to zero, the quicker the sequence converges. (If $g = 0$, then B is at the point of intersection and an exact value is found immediately.)

Alternatively, the equation $x^5 + x - 3 = 0$ can be rearranged to give $x = 3 - x^5$, but when this is solved graphically by looking at where $y = x$ meets $y = 3 - x^5$ you see that the curve has a gradient less than -1 at the point of intersection.

Moving horizontally from the curve to the line and vertically from the line to the curve produces a spiral again, but this time going outwards away from the point whose coordinates you seek.

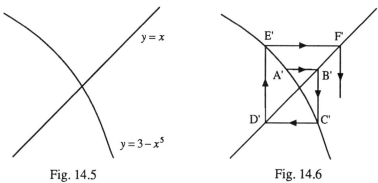

Fig. 14.5   Fig. 14.6

These are the two main cases for curves with **negative** gradients at the required point. ($g = -1$ is a 'square').)

What happens for **positive** gradients?
If the gradient of the curve is between 0 and 1, moving from curve to line and back as before brings you closer and closer the point of intersection on a downward 'staircase'. Again, the closer the gradient is to zero the quicker the convergence will be. (Fig. 14.7)

On the other hand, if the gradient is more than one, a 'staircase' pattern occurs again but this time it leads away from the objective (Fig. 14.8).

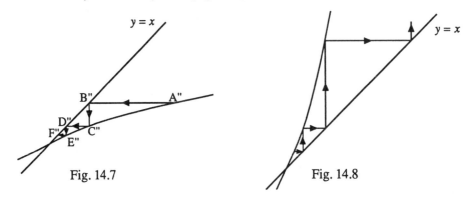

Fig. 14.7   Fig. 14.8

To summarize, if the right-hand side of the iteration formula has a graph which meets $y = x$ at a point where its gradient is between -1 and $+1$, then the associated sequence will converge to the point of intersection as required.

Is it possible to test a formula before applying it?

Yes, it may well be possible to use calculus to find the derivative and then see when it has a value between -1 and $+1$. However, the second equation, (a)(ii), in section 14.2, shows that this may be far from easy. If a sequence diverges then try another rearrangement which may generate something useful.

14.5 ROOTS OF POLYNOMIALS

A quintic may have as many as five solutions, but if

$$ax^5 + bx^4 + cx^3 + dx^2 + ex + f = 0$$

then the solutions add up to $-\dfrac{b}{a}$, so if four are known, the fifth may be calculated from them without the need of any further iteration.

The same principle applies for a quartic, so

$$ax^4 + bx^3 + cx^2 + dx + e = 0$$

has solutions which add up to $-\dfrac{b}{a}$.

For example, $11x^4 + 7x^3 - 6x + 17 = 0$ has solutions which add up to $-\dfrac{7}{11}$ and

$x^3 - 6x - 23 = 0$ has solutions which add up to 0.

Exercise 5.1   What equation is solved by the following iteration formula?

$$x_n = \frac{10}{\sqrt{x_{n-1}}}$$

Use this iteration to solve the equation correct to three decimal places, starting with $x_1 = 1$.

Draw the graphs of $y = x$ and $y = \dfrac{10}{\sqrt{x}}$ for $0 < x \le 10$. Plot the points $(x_1, x_2), (x_2, x_2), (x_2, x_3)$, $(x_3, x_3), (x_3, x_4) \dots (x_7, x_7)$ and join them by straight lines in order.

Exercise 5.2

(a) The cubic equation

$$x^3 - 10x^2 - 2x + 20 = 0$$

may be solved by using the iteration formula

$$x_n = \sqrt{\frac{x_{n-1}^3 - 2x_{n-1} + 20}{10}}$$

Find a solution using this formula and starting with $x_1 = 1$, giving your answer to three decimal places.

(b) Verify that $x = 10$ is also a solution to the equation.

(c) Write down the third solution correct to three decimal places.

Exercise 5.3   Use the iteration formula $x_n = \dfrac{1}{x_{n-1}^2} + 1$ to solve the equation $x^3 - x^2 - 1 = 0$.

Start with $x_1 = 1$ and calculate x_{10}. (Keep x_9 in your calculator memory.) Which of x_9, x_{10} or $\dfrac{x_9 + x_{10}}{2}$ would give the best approximation to the solution? Use this to give a better answer by applying the formula.

Draw the graphs of $y = x$ and $y = \dfrac{1}{x^2} + 1$ for $1 \le x \le 2$. Plot the points $(x_1, x_2), (x_2, x_2)$, $(x_2, x_3), (x_3, x_3), (x_3, x_4) \dots (x_6, x_6)$ and join them in order.

Exercise 5.4

(a) Show that the equation $x^3 - 20x + 8 = 0$ may be arranged to give

$$x = \sqrt{20 - \frac{8}{x}}$$

Use the associated iteration formula with $x_1 = 1$ to find a solution of the original equation correct to four significant figures. Draw graphs of $y = x$ and $y = \sqrt{20 - \dfrac{8}{x}}$ for $0 \le x \le 5$ using 2 cm for each unit in the x-direction and 5 cm per unit in the y-direction. (Note: $y = x$ will not be at $45°$.)

Plot the points $A(x_1, x_2), B(x_2, x_2), C(x_2, x_3), D(x_3, x_3), \dots H(x_5, x_5)$ and join them in order.

(b) Show that the equation $x^3 - 20x + 8 = 0$ may also be arranged to give

$$x = \frac{1}{20}x^3 + 0.4$$

Use the associated iteration formula with $x_1 = 0$ to find a solution to the original equation correct to four significant figures.

(c) Hence estimate the third solution to the equation.

14.6 INTERVAL BISECTION
The idea behind interval bisection is to locate a root of an equation in a particular interval and then repeatedly reduce the length of it.

Example

Solve $f(x) = x^5 + x - 3 = 0$

Solution

$$f(1) = 1 + 1 - 3 = -1$$
$$f(2) = 32 + 2 - 3 = 31$$

Since $f(1) = -1 < 0$ and $f(2) = 31 > 0$, the graph may well cut the x-axis, that is, be zero for some value of x between 1 and 2. To reduce the interval of values in which it may lie try

$$f(1.5) = 7.59375 + 1.5 - 3 = 6.09375 > 0$$

Call the value you are seeking, α.
You know that $1 < \alpha < 2$ and now $1 < \alpha < 1.5$. This is hardly surprising as $f(1)$ was much closer to zero than $f(2)$, so you should expect α to be closer to 1 than 2.
Try 1.2 and subsequent improvements.

$$f(1.2) = 0.68832 > 0$$
$\Rightarrow$ $1 < \alpha < 1.2$

$$f(1.1) = -0.28949 < 0$$
$\Rightarrow$ $1.1 < \alpha < 1.2$

$$f(1.15) = 0.161357187 > 0$$
$\Rightarrow$ $1.1 < \alpha < 1.15$

$$f(1.12) = -0.117658316 < 0$$
$\Rightarrow$ $1.12 < \alpha < 1.15$

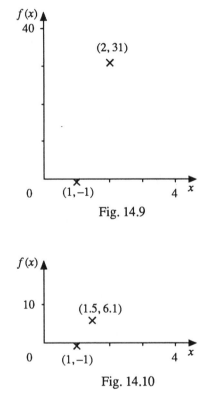

Fig. 14.9

Fig. 14.10

$$f(1.135) = 0.018559343 > 0$$
$$\Rightarrow \quad 1.12 < \alpha < 1.135$$

$$f(1.13) = -0.02756482 < 0$$
$$\Rightarrow \quad 1.13 < \alpha < 1.135$$

$$f(1.133) = 0.00002248939 > 0$$
$$\Rightarrow \quad 1.133 < \alpha < 1.135$$

so $\alpha = 1.13$ to 2 decimal places.

Further investigation is of course possible, to obtain the root to a greater degree of accuracy.

14.7 PROBLEMS WITH INTERVAL BISECTION

(a) Finding the initial interval of values may require a good deal of work. A large number of values may need to be tried before a pair is found so that one produces a positive result and the other a negative value. Perhaps there is a root between –58 and –57; it might take some time for you to identify this.

(b) Trying all the integers between say –20 and 20 may fail to locate any roots even if they exist there. For example,

$$f(x) = 96x^4 - 752x^3 + 1794x^2 - 1243x + 210 = 0$$

has four roots which are $\dfrac{1}{4}, \dfrac{3}{4}, 3\dfrac{1}{3}, 3\dfrac{1}{2}$.

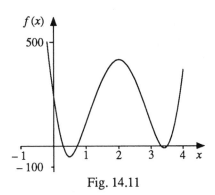

Since

$f(0), f(1), f(2), f(3)$ and $f(4)$
are all positive you can draw no conclusions about the positions of any roots although they all lie in the area you have 'checked'.

Fig. 14.11

(c) The idea of looking for a change of sign may not provide an interval of values even if such a change has been found to occur.

For example, for

$$f(x) = \frac{x-2}{x^3}$$

it is true that

$$f(1) = -1 < 0 \quad \text{and} \quad f(-1) = 3 > 0$$

but it is not true that a root lies between
-1 and 1.   The problem is that there is a
discontinuity between -1 and 1 (Fig.
14.12). Such problems do not occur with
polynomials, but if the function has a
denominator which may become zero,
there will be a discontinuity. The frac-
tion may be concealed as with

$$f(x) = \tan x$$

which is $f(x) = \dfrac{\sin x}{\cos x}$

and $\cos x$ may certainly become zero.

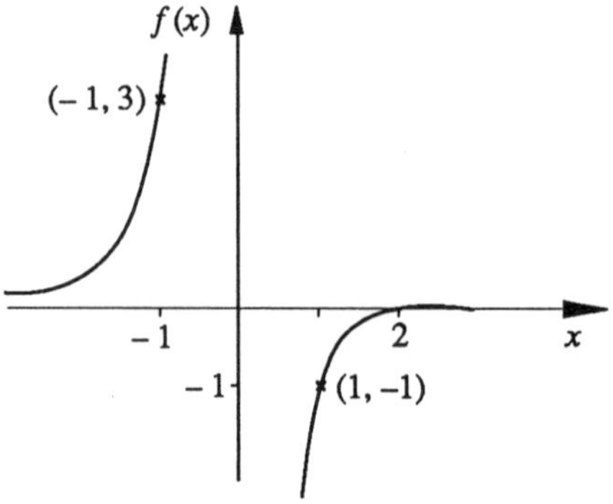

Fig. 14.12

(d) Interval bisection is sometimes tedious and often involves calculations with long
numbers on a calculator which, in turn, can lead to mistakes.

14.8   ITERATION FORMULA v INTERVAL BISECTION

(a) There may be trouble finding an initial interval to bisect whereas if an iteration leads to
convergence, the starting point may not matter.

(b) Once an interval is found, bisection will produce an answer but there is no guarantee that
an iteration formula will provide a convergent sequence.

(c) An initial study of integers may locate several roots which can then be found by
bisection. Iteration gives no idea of the number of roots that exist (so a study of integers
may be useful here too).

In any case, a study of integers which could be used to draw the graph of the function may
prove helpful. Graphic calculators make a sketch of the function immediately available and
their usefulness should not be underestimated.

14.9 HERO'S METHOD FOR SQUARE ROOTS

Consider the iteration formula $x_n = \dfrac{1}{2}\left(x_{n-1} + \dfrac{12}{x_{n-1}} \right)$

If it converges, it does so to the root of the equation

$$x = \frac{1}{2}\left(x + \frac{12}{x} \right)$$

$\Rightarrow \qquad 2x = x + \dfrac{12}{x}$

$\Rightarrow \qquad x = \dfrac{12}{x}$

$\Rightarrow \qquad x^2 = 12$

That is, if it converges at all it does so to $\sqrt{12}$.

Starting with

$$x_1 = 1$$

$$x_2 = \frac{1}{2}(1 + 12) = 6.5$$

$$x_3 = \frac{1}{2}\left(6.5 + \frac{12}{6.5}\right) = 4.173076923$$

$$x_4 = \frac{1}{2}\left(x_3 + \frac{12}{x_3}\right) = 3.52432648$$

$$x_5 = 3.464616186$$

$$x_6 = 3.464101653$$

But $\sqrt{12} = 3.464101615$ so after only five iterations the result is remarkably close to $\sqrt{12}$.

It is not surprising that this method or a similar one is used in modern computers. What is the idea behind it?

Suppose $x_{n-1} < \sqrt{12}$ so the value is too small

$\Rightarrow$ $x_{n-1}\sqrt{12} < 12$ by multiplying by $\sqrt{12}$

$\Rightarrow$ $\sqrt{12} < \dfrac{12}{x_{n-1}}$ so $\dfrac{12}{x_{n-1}}$ is too big

$\Rightarrow$ $x_{n-1} < \sqrt{12} < \dfrac{12}{x_{n-1}}$

Thus a better approximation than x_{n-1} would be the average of x_{n-1} and $\dfrac{12}{x_{n-1}}$, as it is mid-

way between them, that is $\dfrac{1}{2}\left(x_{n-1} + \dfrac{12}{x_{n-1}}\right)$.

Exercise 9.1   Use Hero's method to evaluate the square roots of

(a)   39

(b)   17 starting from a 'suitable' value of x_1. (3 decimal places)

Exercise 9.2   What happens when Hero's method is used to calculate $\sqrt{-39}$?

[i.e. Use $x_n = \dfrac{1}{2}\left(x_{n-1} - \dfrac{39}{x_{n-1}}\right)$]

Exercise 9.3   If the iteration formula $x_n = \dfrac{1}{3}\left(2x_{n-1} + \dfrac{12}{x_{n-1}^2}\right)$ converges, what is the limit of its sequence? Confirm that it does indeed converge.

Exercise 9.4   If the iteration formula $x_n = \dfrac{1}{4}\left(3x_{n-1} + \dfrac{12}{x_{n-1}^3}\right)$ converges, what is the limit of its sequence? Confirm that it does indeed converge.

Exercise 9.5   Write down an iteration formula that you might expect to converge to $\sqrt[9]{12}$.

Exercise 9.6   Use interval bisection to find all three real solutions to

$16x^5 + 16x^4 - x - 1 = 0$ (All three lie between −3 and 3.)

Exercise 9.7   Use interval bisection to find the real root of $x^3 + 3x - 17 = 0$, correct to 3 decimal places.

Exercise 9.8   Find the two solutions of $2^x + x^2 - 31 = 0$ by interval bisection, giving your answers correct to 2 decimal places. (Both answers lie between −6 and +6.)

Exercise 9.9   Use interval bisection to solve $17 - x^x = 0$, correct to 2 decimal places.

14.10 NOTES
Cardano's method

$$x^3 + a^3 + b^3 - 3abx = (x + a + b)\left(x^2 - (a+b)x + (a+b)^2 - 3ab\right)$$

i.e. $x + a + b$ is a factor, $-(a + b)$ is a root.

Replace: $a \to \omega a$ $b \to \omega^2 b$ where $\omega = \sqrt[3]{+1}$

$$x^3 + (a\omega)^3 + (\omega^2 b)^3 - 3(\omega a)(\omega^2 b) = x^3 + a^3 + b^3 - 3ab$$

$\Rightarrow$ $x + a\omega + b\omega^2$ is a factor, $-(a\omega + b\omega^2)$ is a root.

Similarly

     $x + a\omega^2 + b\omega$ is a factor, $-(a\omega^2 + b\omega)$ is a root.

$$x^3 + rx^2 + sx + t = 0 \text{ can be transformed to } x^3 + px + q = 0 \text{ by } x \to x - \frac{r}{3}$$

Consider $x^3 + px + q = 0$

i.e. $p = -3ab, \quad q = a^3 + b^3$ with roots $-a - b, \ -a\omega - b\omega^2, \ -a\omega^2 - b\omega$ (*)

Product of roots $\Rightarrow q = a^3 + b^3$

$$\Sigma \alpha \beta = p \Rightarrow (-a - b)(-a\omega - b\omega^2)$$

$$+(-a - b)(-a\omega^2 - b\omega)$$

$$+(-a\omega - b\omega^2)(-a\omega^2 - b\omega) = p$$

$$\Rightarrow p = a^2\omega + ab\omega + ab\omega^2 + b^2\omega^2 + a^2\omega^2 + ab\omega^2 + ab\omega$$

$$+ b^2\omega + a^2\omega^3 + b^2\omega^3 + ab\omega^2 + ab\omega$$

$$= a^2(\omega + \omega^2 + \omega^3) + b^2(\omega + \omega^2 + \omega^3) + 3ab(\omega + \omega^2)$$

$$= -3ab \text{ as sum of roots of } x^3 - 1 = 0 \text{ is zero, i.e. } \omega + \omega^2 + 1 = 0.$$

i.e. $p = -3ab$ and $q = a^3 + b^3$

a^3, b^3 are roots of $y^2 - qy - \dfrac{p^3}{27} = 0$

$$\Rightarrow a^3 = \frac{q \pm \sqrt{q^2 - 4 \times \left(-\dfrac{p^3}{27}\right)}}{2} \qquad [\text{or } b^3 =]$$

$$\Rightarrow a^3 = \frac{q}{2} + \sqrt{\frac{q^2}{4} + \frac{p^3}{27}} \qquad b^3 = \frac{q}{2} - \sqrt{\frac{q^2}{4} + \frac{p^3}{27}}$$

$$\Rightarrow a = \left(\frac{q}{2} + \sqrt{\frac{q^2}{4} + \frac{p^3}{27}}\right)^{\frac{1}{3}} \qquad b = \left(\frac{q}{2} - \sqrt{\frac{q^2}{4} + \frac{p^3}{27}}\right)^{\frac{1}{3}}$$

This gives the 3 roots of (*), but the nature of the roots is decided by the discriminant

$$D = \frac{q^2}{4} + \frac{p^3}{27}$$

Case 1: $D > 0 \implies a$ and b are both real, so the only real root is $-a-b$. The other

roots $\left(-a\omega - b\omega^2, -a\omega^2 - b\omega\right)$ are complex.

Case 2: $D = 0$ $a = b = \sqrt[3]{\dfrac{q}{2}}$, so the roots are

$$-a - b = -2a = -2\sqrt[3]{\frac{q}{2}}$$

$$-a\omega - b\omega^2 = -a\omega - a\omega^2 = a = \sqrt[3]{\frac{q}{2}}$$

$$-a\omega^2 - b\omega = -a\omega^2 - a\omega = a = \sqrt[3]{\frac{q}{2}}$$

Case 3: $D < 0 \implies a^3, b^3$ are complex so De Moivre's theorem is used to find a and b.

$a^* = b$ so, for example,

$$a = u + iv \qquad b = u - iv$$

and the roots are

$$-a - b = -2u$$

$$-a\omega - b\omega^2 = -\omega(u + iv) - \omega^2(u - iv)$$

$$= \left(-\omega - \omega^2\right)u + v\left(i\omega^2 - i\omega\right)$$

$$= u + vi\left(\frac{-1 + \sqrt{3}\,i}{2} - \frac{-1 - \sqrt{3}\,i}{2}\right)$$

$$= u - v\sqrt{3}$$

$$-a\omega^2 - b\omega = u + v\sqrt{3}$$

15

Sorting and packing

15.1 INVESTIGATION

Six people (represented by the numbers 1 to 6) of different heights are standing in a row.

4   6   1   3   5   2

The shortest is number one, the second shortest is number two, and so on.

(a) Arrange them in a line so that they are in ascending order of height with the shortest on the left, subject to the rule that the only move allowed is to swap adjacent people in the row.

(b) What is the least number of moves with which you can obtain the correct order?

(c) Do your moves conform to a pattern? If so, write down the rule that you have developed to produce the moves.

15.2 SORTING

There are many **sorting algorithms** available; some are more efficient than others. In some cases the advantage of using one algorithm as opposed to another may depend on the data to be sorted, so it may be worth studying the problem before choosing a particular method.

Note: It is worth remembering that developments and refinements are continually being made to sorting algorithms, so different texts may give procedures under the same heading which are not identical. This will certainly be true of shell sorts and quick sorts.

With the addition of line numbers, the skeleton programs given will run on most PCs.

15.3 BUBBLE SORTING

A **bubble sort** works by passing through the data placing the last element (tallest person) in the correct place (at the extreme right of the row), by comparing pairs. A second run through the data then puts the penultimate element in its correct position, and so on. Just as a bubble rises to the surface of liquid, so the tallest moves to the end.

This can be illustrated by seeing how it works on the row of six people you started with. The pairs being considered are enclosed in a rectangle. If the comparison results in an exchange, the relevant numbers have an arrowed line between them.

$$\boxed{4 \quad 6} \quad 1 \quad 3 \quad 5 \quad 2$$

$$4 \quad 6{\longleftrightarrow}1 \quad 3 \quad 5 \quad 2$$

The first pass through the data looks at 6 elements and so makes 5 comparisons.

$$4 \quad 1 \quad 6{\longleftrightarrow}3 \quad 5 \quad 2$$

$$4 \quad 1 \quad 3 \quad 6{\longleftrightarrow}5 \quad 2$$

$$4 \quad 1 \quad 3 \quad 5 \quad 6{\longleftrightarrow}2$$

$$4 \quad 1 \quad 3 \quad 5 \quad 2 \quad 6$$

First pass

Now there are only 5 elements to sort so the method is recursive.

$$4{\longleftrightarrow}1 \quad 3 \quad 5 \quad 2 \quad 6$$

$$1 \quad 4{\longleftrightarrow}3 \quad 5 \quad 2 \quad 6$$

The second pass may again look at 5 comparisons. When 5 passes have been made, 5 elements are in the correct place so the sixth must be also.

$$1 \quad 3 \quad 4{\longleftrightarrow}5 \quad 2 \quad 6$$

$$1 \quad 3 \quad 4 \quad 5{\longleftrightarrow}2 \quad 6$$

$$1 \quad 3 \quad 4 \quad 2 \quad \boxed{5 \quad 6}$$

$$1 \quad 3 \quad 4 \quad 2 \quad 5 \quad 6$$

Second pass

Six passes would produce 6×5 comparisons. The last pass may be omitted, giving 5×5 comparisons. The fifth comparison in the second pass may be left out as it is irrelevant. Similarly, the third pass may be reduced by two comparisons as only three are essential. The number of comparisons is then

$$5 + 4 + 3 + 2 + 1 = 15.$$

In general, the number of comparisons required is

$$(n-1)+(n-2)+...+1 = \frac{n(n-1)}{2}$$

where n is the number of elements to be sorted.

Continue the process on the six people and note the number of exchanges made.

Skeleton bubble sort

This program is written in Quick BASIC 4.5 and should run on most PCs. Introductory lines can be found in section 15.7.

```
                          'Sort done on the array Num () having n elements
BUBBLESORT:                           'Label beginning line of routine
flag = 0                              'Zero an exchange indicator
FOR x = 1 TO n - 1                    'Step through array (less one)
   IF Num(x) > Num(x + 1) THEN 'Is exchange necessary? If so then
                                      'swop them by doing –

      Store = Num(x)                     'Save first element in Store
      Num(x) = Num(x + 1)                'Make the two elements equal
      Num(x + 1) = STORE                 'Make second element equal Store
      flag = -1                          'And show exchange has been made
   END IF
NEXT                                  'Move on to next element in array
If flag = -1 THEN                     'If at least one exchange has been done
GOTO BUBBLESORT                       'then repeat whole process
END IF
RETURN                                'Go back to original place in program
```

15.4 SHUTTLE SORT

The **shuttle sort** compares the two elements at the start and puts them in the correct order. The next element is introduced, compared with those already sorted and shuttled along the line to its correct position.

How does this operate on the six people in the row you have been considering ?
As before, comparisons are shown by rectangles and exchanges by ⟷ . Initially only two elements are compared and exchanged if necessary.

Those considered are to the left of the dotted line.

| 4 | 6 | 1 | 3 | 5 | 2 |

The third element in the row is brought into consideration and shuttled along to its correct position so that those to the left of the dotted line are in the correct order.

4   6⟷1   ¦3   5   2

4⟷1   6   ¦3   5   2

1   4   6   ¦3   5   2

1	4	6⟷3	⦙ 5	2

The 3 is shuttled along to its correct position in the set of four.

1	4⟷3	6	⦙ 5	2

| 1 | 3 | 4 | 6 | ⦙ 5 | 2 |

| 1 | 3 | 4 | 6 | ⦙ 5 | 2 |

Only two comparisons are necessary to put the 5 in the correct place in the current list of five elements.

| 1 | 3 | 4 | 6⟷5 | ⦙ 2 |

| 1 | 3 | 4 | 5 | 6 | ⦙ 2 |

| 1 | 3 | 4 | 5 | 6 | ⦙ 2 |

The first pass uses just one comparison. The second pass needs two and the third employs three. If this is continued it produces

$$1 + 2 + 3 + 4 + 5 = 15 \text{ comparisons, and in general}$$

$$1 + 2 + 3 + \ldots + (n - 1) = \frac{n(n - 1)}{2} \text{ comparisons.}$$

As can be seen from the fourth pass, however, the number of comparisons can be reduced below this.

Skeleton shuttle sort

This program is written in Quick BASIC 4.5 and should run on most PCs. Introductory lines can be found in section 15.7.

```
SHUTTLESORT:                        'Label needed for main program
FOR x = 1 TO n - 1                  Step through array.
  IF Num(x) > Num(x + 1)  THEN      Is exchange necessary?
     Store = Num(x + 1)             Yes; so swap elements and take note
                                    of element to be repositioned and
                                    look back down the list for its correct
                                    position.

     Num(x + 1) = Num(x)

     y = x                          y now marks correct position so room
                                    must be made for it by pushing
                                    elements above up one place.

LOOKBACK Shuttle:
    IF y > 1 THEN
      IF Num(y - 1) > Store THEN
```

```
        y = y - 1
        GOTO LOOKBACK SHUTTLE
     END IF
   END IF
   IF x > y THEN
      FOR z = x - 1 TO y STEP -1
         Num(z + 1) = Num(z)
      NEXT
   END IF
   Num(y) = Store              Put element into correct position.
  END IF
NEXT
RETURN
```

15.5 SHELL SORT

The **shell sort** differs from the bubble and shuttle methods as it compares, and possibly exchanges, non-adjacent elements. The set of elements is split into **subsets**. The number of subsets for the first pass is $INT\left(\dfrac{n}{2}\right)$, that is, the number of elements, divided by two and ignoring any remainder.

For this set   4   6   1   3   5   2

you have $INT\left(\dfrac{6}{2}\right) = 3$, so there are three subsets in the first pass. The three subsets are: the first and fourth elements; the second and fifth elements; the third and sixth elements. These pairs are shuttle sorted.

	A	B	C	A	B	C
Subset				.		
	4	6	1	3	5	2
	3	5	1	4	6	2

The 3 and 4 are exchanged, as are the 5 and 6.

For the second pass there is $INT\left(\dfrac{3}{2}\right)$ that is just one subset. A shuttle sort is now carried out on the list.

Shuttle on two elements.

$$\boxed{3 \quad\quad 5} \quad 1 \quad\quad 4 \quad\quad 6 \quad\quad 2$$

$$3 \quad\quad 5{\longleftrightarrow}1 \ \vdots \ 4 \quad\quad 6 \quad\quad 2$$

Introduce the 1 and shuttle sort three elements.

$$3{\longleftrightarrow}1 \quad\quad 5 \ \vdots \ 4 \quad\quad 6 \quad\quad 2$$

Shuttle the 4 until it reaches the correct position.   1   3   5⟵⟶4   ┆6   2

                                           1 | 3   4 | 5   ┆6   2

                                             1   3   4 | 5   6 | ┆2

The final pass puts the 2 in the correct place in the list.
                                 1   3   4   5   6⟵⟶2

                                 1   3   4   5⟵⟶2   6

                                 1   3   4⟵⟶2   5   6

                                 1   3⟵⟶2   4   5   6

                                 1   2   3   4   5   6

Now see how the method works on a larger list.

        4   7   6   1   9   8   3   5   2

The number of subsets in the first pass is $\text{INT}\left(\dfrac{9}{2}\right) = 4$.

These are $\{4, 9, 2\}, \{7, 8\}, \{6, 3\}, \{1, 5\}$.

        A   B   C   D   A   B   C   D   A

        4   7   6   1   9   8   3   5   2

These subsets A, B, C and D are shuttle sorted to produce

        2   7   3   1   4   8   6   5   9

The number of subsets in the second pass is $\text{INT}\left(\dfrac{4}{2}\right) = 2$.

        A   B   A   B   A   B   A   B   A

        2   7   3   1   4   8   6   5   9

These two subsets are now shuttle sorted to give

| 2 | 1 | 3 | 5 | 4 | 7 | 6 | 8 | 9 |

Since $A = \{2, 3, 4, 6, 9\}$ and $B = \{7, 1, 8, 5\}$, the elements of A have remained unmoved as they were correctly placed within that subset, but B has been reordered.

7	1	8	5
	⇓		
1	5	7	8

For the third pass the number of subsets is $\mathrm{INT}\left(\dfrac{2}{2}\right) = 1$, so

| 2 | 1 | 3 | 5 | 4 | 7 | 6 | 8 | 9 |

is shuttle sorted to produce

| 1 | 2 | 3 | 4 | 5 | 6 | 7 | 8 | 9 |

Notice how close the result of the second pass is to the desired arrangement.

At each stage,

$$\text{number of subsets in pass} = \mathrm{INT}\left(\frac{\text{number of subsets in last pass}}{2}\right)$$

Skeleton shell sort

This program is written in Quick BASIC 4.5 and should run on most PCs. Introductory lines can be found in section 15.7.

```
SORT:
s = INT (n/2)
```
Number of subsets is $\mathrm{INT}\left(\dfrac{n}{2}\right)$ begin at the start of the array.

```
NEW SORT:
e = 1
NEW ELEMENT:
i = e
```
Sort at this position.

```
PASS ON:
j = i + s
IF a$(i) > a$(j) THEN
```
j corresponds with i's position.

Compare elements and swap via a$ if necessary.

```
a$ = a$(j)
```

```
a$(j) = a$(i)
a$(i) = a$
i = i - s                              Reposition swapped elements.
IF i > 0 THEN GOTO PASS ON.
END IF
e = e + 1                              Move to next subset. Compare element
                                       with the next in the subset, unless this is
                                       the last. Reduce the number of subsets.
IF e < = 1 THEN GOTO NEW ELEMENT
s = INT (S/2)                          If not all sorted then re-sort with new
                                       subsets.

IF s > 0 THEN
   GOTO NEW SORT
END IF
RETURN                                 Sort is complete.
```

15.6 QUICK SORT (SUPER–POINTER VERSION)

The quick sort has two **pointers** indicating elements in the list. The symbol ↓ is used as the stationary pointer and | as the mobile pointer.

The pointers start at opposite ends of the array and the two elements indicated are compared to see if they need to be swapped. If they do, then they are interchanged and carry their pointers with them.

$$
\begin{array}{cccccc}
\downarrow & & & & & | \\
4 & 6 & 1 & 3 & 5 & 2
\end{array}
$$

(The pointers can start at either end.)

$$
\begin{array}{cccccc}
| & & & & & \downarrow \\
2 & 6 & 1 & 3 & 5 & 4
\end{array}
$$

The mobile pointer then moves one place towards the stationary one and the indicated elements are compared again. This process continues until the two pointers meet.

If there is more than one element above the pointers then that subset is quick sorted; likewise for any subset below the pointers. The method can, therefore, be seen as recursive.

$$
\begin{array}{cccccc}
 & | & & & & \downarrow \\
2 & 6 & 1 & 3 & 5 & 4
\end{array}
$$

$$
\begin{array}{cccccc}
 & \downarrow & & & & | \\
2 & 4 & 1 & 3 & 5 & 6
\end{array}
$$

$$
\begin{array}{cccccc}
 & \downarrow & & & | & \\
2 & 4 & 1 & 3 & 5 & 6
\end{array}
$$

$$
\begin{array}{cccccc}
 & \downarrow & & | & & \\
2 & 4 & 1 & 3 & 5 & 6
\end{array}
$$

$$
\begin{array}{cccccc}
 & | & & \downarrow & & \\
2 & 3 & 1 & 4 & 5 & 6
\end{array}
$$

$$
\begin{array}{cccccc}
 & & | & \downarrow & & \\
2 & 3 & 1 & 4 & 5 & 6
\end{array}
$$

With only three exchanges made the list is looking close to the desired 1 2 3 4 5 6 order.

With n elements the quick sort may require of the order of n^2 comparisons and exchanges but this is unlikely to occur. In practice the quick sort is, as it its name suggests, best for handling large arrays and can save a considerable time when compared with, for example, the bubble sort.

Skeleton quick sort

```
SORT:
    tp$ = ","                          Clear top pointer stack.
    b $ = " "                          Clear bottom pointer stack.
    tp = 1                             Set the pointers at either end of the
    bp = n                             array.
    NEXT SORT:
    ti = tp                            Store initial value of both pointers.
    bi = bp
    NEW ELEMENT:
IF bp > tp THEN                        Ensure that t and b hold the top and
    t = tp                             bottom pointers the correct way round.
    b = bp
    ELSE
    t = bp
    b = tp
END IF
IFa$(t) > a$(b) THEN                   If exchange is necessary then
    a$ = a$ (bp)                       swap elements.
    a$(bp) = a$(tp)
    a$(tp) = a$
    p = bp                             Swap pointers.
    bp = tp
    tp = p
END IF
IFtp < >bp THEN                        Move bottom pointer towards top
    bp = bp + SGN(tp-bp)               pointer.
    GOTO NEW ELEMENT                   Examine next pair of elements.
END IF
IFtp - ti > 1 THEN
    tp$ = tp $ + STR$(ti) + ""
    bp$ = bp $ + STR$(tp - 1) + ","
END IF
IFbi - bp > 1 THEN                     If bottom sublist exists, add it to the
```

```
  tp$ = tp $ + STR$(bp+1) + ""            pointer stacks.
  bp$ = bp $ + STR$(bi) + ","
END IF
IFtp$ = "," THEN RETURN                    All sublists now sorted.
tp = VAL(RIGHT$(tp$, LEN(tp$)-1)          Extract new pointer
b p = VAL(RIGHT$(bp$, LEN(bp$)-1)         values from both stacks.
STRIP TOP:
tp$ = RIGHT$(tp$, LEN(tp$)-1)             Erase new pointer values
IFLEFT $(tp$, 1)< > "," THEN              from both stacks.
    GOTO STRIP TOP
END IF
STRIP BOT:
bp$ = RIGHT$(bp$, LEN(bp$)-1)
IFLEFT$(bp$, 1) < > "," THEN
    GOTO STRIP BOT
END IF
GOTO NEXT SORT                             Sort new sublist.
```

15.7   FIRST HALF OF SORT PROGRAM

```
CLS
  NEXT LIST                                Clear all variables.
  CLEAR
  INPUT "Number of items to SORT:, n      See note below.
  IF n = 0 THEN END
  PRINT
  DIM a$ (1 to n)
FOR= 1 to n
    PRINT "Type in item"; a;
    INPUT " : " , a$ (a)
  NEXT
  PRINT                                    Perform sort.
  PRINT "Sorting ...."
  GOSUB SORT
  PRINT
  PRINT "SORTED LIST"
  PRINT "--------------------"
  PRINT                                    Print out sorted array.
```

```
FOR a = 1 TO n
   PRINT a$(a)                        Prompt for new array.
NEXT
PRINT
     PRINT
GOTO NEXT LIST
```

Note: Array is dimensioned from 1 to n for clarity. The simplest idea may be to replace the DIM with DIM a$(n), leaving a$(0) existent but unused.

Exercise 7.1 Use the bubble, shuttle and shell sorts on the following sets of data, noting the numbers of comparisons and exchanges for each.

(a)   8   7   6   5   4   3   2   1

(b)   2   3   4   5   6   7   8   1

Exercise 7.2 A computer takes one unit of time to make a comparison and two units to complete an exchange. Use the formula $T = C + 2E$ where T = total time, E = number of exchanges and C = number of comparisons, to comment on the times taken in Exercise 7.1.

Exercise 7.3 Use the quick sort and bubble sort methods on the following sets of data, noting the numbers of comparisons and exchanges.

(a)   77   22   8   11   25

(b)   3   17   12   9   2   11   7

Comment on your results.

Exercise 7.4 Use the formula $T = C + 2E$ to compare the performances of the shell and quick sorts on

                7   11   4   9   2   5   6

Exercise 7.5
(a) What is the principle difference in the methods by which the bubble sort and the shuttle sort deal with incorrectly positioned elements?
(b) Once the bubble sort has examined the last element of an array for the first time, this does not mean that the sort is complete. However, after the shuttle sort has dealt with the last element of an array, this does mean that the sort is complete. What is the reason for this difference?

Exercise 7.6 The following array is sorted into ascending order by the shell sort.

position	1	2	3	4	5	6	7	8
element	2	2	2	2	2	2	2	1

Which positions will contain the number 1 at any time?

Exercise 7.7 How can the idea of recursion be applied to the quick sort algorithm?

Exercise 7.8 In the worst-case scenario, the time required for the quick sort to sort n elements is of the order n^2. The bubble sort also requires a time of the order of n^2 to sort an equivalent array of elements. With this in mind, why is the quick sort considered so much more suitable than the bubble sort for sorting large arrays?

15.8 (BIN) PACKING
Investigation
Ali, Chris and Jo are waiting for the lorry bringing the final delivery of the day to the warehouse where they work. They know that the following 14 loads are on board.

load	A	B	C	D	E	F	G	H	I	J	K	L	M	N
weight	13	5	14	11	2	4	4	10	5	3	8	7	9	5

(weights in 100s of kg)

The lift from their unloading bay can handle a maximum of 2000 kg at a time. In order to leave work as early as possible they want to use the minimum number of journeys in the lift.

What is the most efficient way the lift can be loaded each time?

The total delivery is

$$13 + 5 + 14 + 11 + 2 + 4 + 4 + 10 + 5 + 3 + 8 + 7 + 9 + 5$$

$$= 100 \times 100 \text{ kg}$$

$$= 10\,000 \text{ kg}$$

Can it all be moved by the lift in five trips?

What is the best possible arrangement?

15.9 PACKING ALGORITHMS
First fit algorithm
It is decided that Ali will start with load A, assigning it to trip 1 in the lift, and then working through the loads in the order listed, fitting each one into the first available lift journey with sufficient space. This will produce the arrangement in Fig. 15.1, and the whole delivery can be moved in 6 trips.

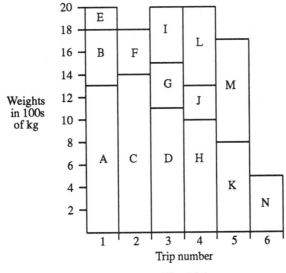

Fig. 15.1

Decreasing first fit algorithm

Chris realises that it might be best to decide at the outset where to put the large loads. For example, it is better to be left with a 300 kg load and a 200 kg load than with a single load of 500 kg (as Ali discovered). Chris puts the list in descending order of weight and then uses the same idea of putting each load into the first available space big enough. The result is shown in Fig. 15.2.

load	C	A	D	H	M	K	L	B	I	N	F	G	J	E
weight	14	13	11	10	9	8	7	5	5	5	4	4	3	2

(weights in 100s of kg)

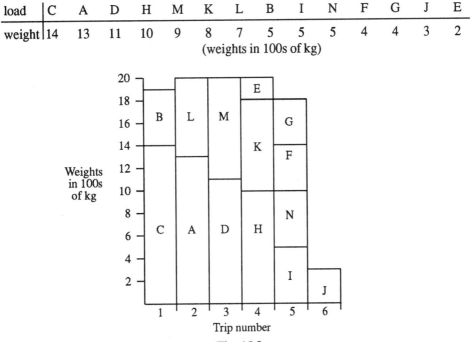

Fig. 15.2

Perhaps Chris' method (Fig. 15.2) is better as less is left for the final trip, but again six lift journeys are needed.

Brute force / exhaustion algorithm

Jo decides that the best plan is to ensure that the lift is full each time and so to look for combinations that total 20. The first such combination is

$$A + B + E = 13 + 5 + 2 = 20$$

Now the problem is reduced to fitting the remaining loads,

load	C	D	F	G	H	I	J	K	L	M	N
weight	14	11	4	4	10	5	3	8	7	9	5

(weights in 100s of kg)

in to a minimum number of lift journeys, preferably four. Already there is a problem as the 14 cannot be complemented to give a total of 20, so at least six trips will be needed.

Now $D + F + I = 11 + 4 + 5 = 20$

then $H + J + L = 10 + 3 + 7 = 20$

No obvious combinations remain so Jo decides to undo the last combination, and tries

$$G + L + M = 4 + 7 + 9 = 20$$

Once again, there is no way of combining the remaining loads into 2 lift trips. Unfortunately, the methods used do not lead to the best solution, which involves shifting the load in five trips. Fig. 15.3 shows a possible solution using five trips.

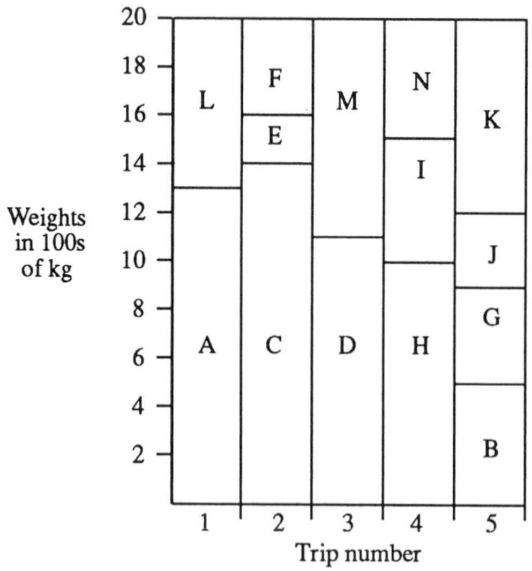

Fig. 15.3

This solution is not unique, as M may be interchanged with B and G; also F and G as well as B, I and N, can be swapped amongst themselves.

Given enough time, Jo would have found a solution to the problem.

Of course, larger numbers of objects and 'lifts' would make the process very laborious, particularly as checking all the possible combinations may simply reveal that no ideal arrangement exists.

Ali's algorithm will produce a way to move the goods with little calculation and so is to be applauded for speed and simplicity. However, Chris's idea that the large objects should be placed first, as smaller ones are easier to deal with at the end, is likely to find the optimal arrangement, or to produce an answer closer to the ideal than Ali's.

How would Ali, Chris and Jo have fared with the following loads?

load	A	B	C	D	E	F	G	H	I	J	K	L	M
weight	11	16	4	6	2	8	6	9	13	14	6	1	4

(weights in 100s of kg)

Ali would quickly arrive at the arrangement shown in Fig. 15.4, so these loads can be despatched in six batches.

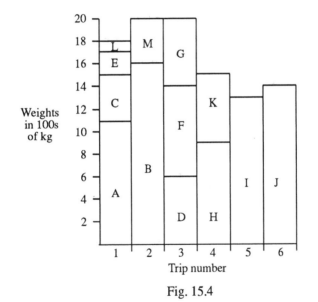

Fig. 15.4

Chris, on the other hand, would sort the loads into order.

load	B	J	I	A	H	F	D	G	K	C	M	E	L
weight	16	14	13	11	9	8	6	6	6	4	4	2	1

(weights in 100s of kg)

This would result in the arrangement in Fig. 15.5.

Jo might arrive at a solution quickly, or still be trying some time after the others have gone home!

In each of these examples of loads to be arranged, the total load will fit exactly into multiples of 20, so, with the ideal solution, the lift will be full each time. How would you deal with a delivery of 97 (100 kg)?

This can be done just as before, but those writing a program to do the job may find it useful to introduce three dummy loads, each of 1 unit. The aim is to fill each lift; whether it is full or not may be easily checked.

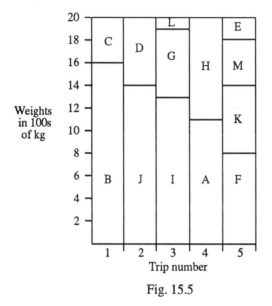

Fig. 15.5

Exercise 9.1   A secondary school is due to receive pupils from its feeder schools as follows.

school	A	B	C	D	E	F	G	H	I	J
no. of pupils	6	24	9	17	4	22	8	7	16	15

The pupils are to be assigned to four classes, each of 32 pupils, in such a way that those from the same feeder school are kept together. If this is possible, show how it can be done.

Exercise 9.2   A carpenter requires the following lengths of wood for a job.

piece	A	B	C	D	E	F	G	H	I	J
length (cm)	75	200	100	175	75	50	125	175	75	150

How many 3 metre lengths will be needed?

Exercise 9.3   A project requires 13 separate tasks to be completed. If five people are each available to work for eight hours a day, can the project be finished in two days?

task	A	B	C	D	E	F	G	H	I	J	K	L	M
time (hrs)	9	3	8	6	5	9	4	6	6	11	5	4	3

Exercise 9.4   During the Christmas period, Bobi wants to record 13 television programmes on six 3-hour video tapes. The lengths of the programmes are as given in the table. Can this be done?

programme	A	B	C	D	E	F	G	H	I	J	K	L	M
duration (mins)	150	90	75	75	45	30	60	150	105	45	105	75	75

Exercise 9.5   During the day, trains will drop off trucks at a station which has six empty sidings, each of which can take 12 trucks. For ease of shunting, the plan is to keep trucks in the groups in which they arrive. Show how this may be done.

train	A	B	C	D	E	F	G	H	I	J	K	L	M	N
no. of trucks	9	7	4	5	8	3	4	1	6	5	5	3	4	8

Exercise 9.6   A school trip is organised to take 90 pupils to a theatre. Six rows of 15 seats have been booked. Show how the pupils may be seated in order to keep pupils in the same form together in the same row.

form	A	B	C	D	E	F	G	H	I	J	K	L	M	N
no. of pupils	6	7	6	10	5	9	12	2	3	4	8	10	3	5

15.10 NOTES

There are many other sorts available to anyone who wants to study the subject further. Some are variations on those you have seen here. They include the *exchange sort, heap sort, insertion sort, merge sort, scatter sort* and *tournament sort*.

It is quite possible to use simple ideas as the bases of sorts. For example, find the mean

of $\dfrac{4+6+1+3+5+2}{6} = 3.5$ (the numbers to be sorted) and compare each element with

the mean. As elements are found which are less than or equal to the mean, they are put at the front of the list and as others are found that are greater than the mean, these are put in the nth, $(n-1)$th, $(n-2)$th places, etc. Now the two sublists can be sorted by any of the methods already discussed or can be 'mean sorted' again.

16

Algorithms

16.1 WHAT IS AN ALGORITHM?

An **algorithm** is a set of steps to carry out a calculation or solve a problem.

- An algorithm must be capable of receiving data.

- An algorithm must be applicable to any appropriate set of data.

- Each step must be exactly defined.

- An algorithm must produce output.

- Its output must be produced after a finite (though possibly very large) number of steps.

Example

Produce a sequence of ten numbers using the following algorithms, (a), (b) and (c).

(a) Algorithm written in English

    Step 1   The first term is 1.

    Step 2   Multiply the last term by 2, add 1 and write down the result.

    Step 3   Repeat step 2 until you have 10 terms, then stop.

(b) Algorithm given as a flow diagram

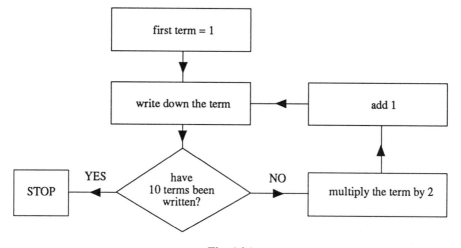

Fig. 16.1

(c) Algorithm given in computer language

Step 1: **t = 1**	first term
Step 2: **n = 1**	counter
Step 3: **PRINT t**	print the term
Step 4: **t = 2*t + 1**	produce new term
Step 5: **n = n + 1**	change counter
Step 6: **IF n ≤ 10 THEN GOTO STEP 3**	10 printed? No, repeat process
Step 7: **STOP**	Yes, stop

Rather than using a single letter for each variable, this may also be written

Step 1: **term = 1**
Step 2: **counter = 1**
Step 3: **PRINT term**
Step 4: **term = 2*term + 1**
Step 5: **counter = counter + 1**
Step 6: **IF counter ≤ 10 THEN GOTO STEP 3**
Step 7: **STOP**

Note the need for a stopping condition. In generating sequences and other processes a step designed to terminate the work may be necessary. Sorting algorithms may stop when the list has been worked through but an iterative process to solve an equation may go on *ad infinitum.*

16.2 INVESTIGATIONS

1. Long division is an example of an algorithm used in calculation.

$$
\begin{array}{r}
8\ 5 \\
14\,\overline{)1\ 2\ 0\ 0\ 2\ 2} \\
1\ 1\ 2 \\
\hline
8\ 0 \\
7\ 0 \\
\hline
1\ 0\ 2 \\
\text{etc.}
\end{array}
$$

The same set of steps is used repeatedly until the division is completed or an answer is obtained to the required degree of accuracy. Write out, in English, an algorithm to divide one positive integer by another.

2. Square roots can be calculated in a similar manner. An example of such a calculation is shown here in various stages.

$$\sqrt{6\,|\,1\ 8\,|\,5\ 1\,|\,6\ 9\,|}$$

$$
\begin{array}{l}
\quad 2 \\
\ \sqrt{6\,|\,1\ 8\,|\,5\ 1\,|\,6\ 9\,|} \\
2\quad 4 \\
\quad \overline{2}
\end{array}
$$

$$
\begin{array}{l}
\quad 2\quad ? \\
\ \sqrt{6\,|\,1\ 8\,|\,5\ 1\,|\,6\ 9\,|} \\
2\quad 4 \\
\quad \overline{2\ 1\ 8} \\
4\,?
\end{array}
$$

$$
\begin{array}{l}
\quad 2\quad 4 \\
\ \sqrt{6\,|\,1\ 8\,|\,5\ 1\,|\,6\ 9\,|} \\
2\quad 4 \\
\quad \overline{2\ 1\ 8} \\
4\,4\quad 1\ 7\ 6 \\
\quad \overline{\ \ 4\ 2}
\end{array}
$$

$$
\begin{array}{l}
\quad 2\quad 4\quad ? \\
\ \sqrt{6\,|\,1\ 8\,|\,5\ 1\,|\,6\ 9\,|} \\
2\quad 4 \\
\quad \overline{2\ 1\ 8} \\
4\,4\quad 1\ 7\ 6 \\
\quad \overline{\ \ 4\ 2\ 5\ 1} \\
4\,8\,?
\end{array}
$$

```
            2   4   8
         _____
       √  6¦1 8¦5 1¦6 9¦
  2       4
          ____
          2 1 8
 44       1 7 6
          _____
          4 2 5 1
488       3 9 0 4
          _____
            3 4 7
```

```
            2   4   8   ?
         _____
       √  6¦1 8¦5 1¦6 9¦
  2       4
          ____
          2 1 8
 44       1 7 6
          _____
          4 2 5 1
488       3 9 0 4
          _____
            3 4 7 6 9
496?
```

```
            2   4   8   7
         _____
       √  6¦1 8¦5 1¦6 9¦
  2       4
          ____
          2 1 8
 44       1 7 6
          _____
          4 2 5 1
488       3 9 0 4
          _____
            3 4 7 6 9
4967        3 4 7 6 9
            _____
                  0
```

Study how the method works, then use it to find

(a) $\sqrt{656076996}$ (b) $\sqrt{8122876129}$

How could the method be used to calculate $\sqrt{3}$?

Explain to someone else how this method works.

Write an algorithm to calculate square roots based on this method.

16.3 SIMPLE ALGORITHMS
Example 1
The following example works through a list of numbers counting the number of fives it contains.

The list of numbers is denoted by

$S(1), S(2), S(3), ..., S(N)$

and therefore has N numbers.

Step

```
1    counter = 0
2    INPUT N
3    FOR X = 1 TO N
4    IF S(X) = 5 THEN counter = counter +1
5    STOP
```

For the list 3, 5, 7, 4, 8, 5, that is $S(1)= 3$, $S(2) = 5$, $S(3) = 7$, $S(4)= 4$, $S(5) = 8$, $S(6) = 5$, the algorithm works as follows:

```
1    counter = 0
2    INPUT N = 6              (there are 6 numbers to check)
3    X = 1
4    S(1) = 3 ≠ 5
3    X = 2
4    S(2) = 5 so counter = 0 + 1 = 1
3    X = 3
4    S(3) = 7 ≠ 5
3    X = 4
4    S(4) = 4 ≠ 5
3    X = 5
4    S(5) = 8 ≠ 5
3    X = 6
4    S(6) = 5 so counter = 1 + 1 = 2
```

The counter now holds the value 2 which is the number of fives. To turn this into a computer program several minor changes would be needed and the lines

```
NEXT X
PRINT counter
```

or similar would be needed between steps 4 and 5.

Example 2

In the following example, $|A - B|$ means the size of $A - B$, so $|6 - 4| = 2$, $|1 - 7| = 6$. The algorithm described by the flow diagram in Fig. 16.2 produces the highest common factor of two numbers, A and B.

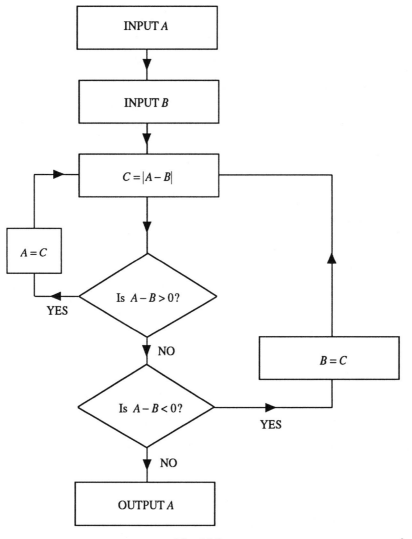

Fig. 16.2

This would work on the numbers 126 and 72 as follows.

Input $A = 126$

Input $B = 72$ A B C

$C = |\,126 - 72\,| = 54$ 126   72   54

$A - B = 126 - 72 > 0?$ Yes

$A = 54$ 54   72   54

$C = |\,54 - 72\,| = 18$ 54   72   18

$A - B = -18 > 0?$ No

$A - B = -18 < 0?$ Yes

	A	B	C
$B = 18$	54	18	18
$C = \left| 54 - 18 \right| = 36$	54	18	36
$A - B = 36 > 0$? Yes			
$A = 36$	36	18	36
$C = \left| 36 - 18 \right| = 18$	36	18	36
$A - B = 18 > 0$? Yes			
$A = 18$	18	18	18
$C = \left| 18 - 18 \right| = 0$	18	18	0
$A - B = 0 > 0$? No			
$A - B = 0 < 0$? No			
Output $= 18$			

At each stage the algorithm replaces the larger of the two numbers by the difference of the two. Since the H.C.F. goes into both numbers it is also a factor of their difference and is the highest one.

Exercise 3.1   Write an algorithm in the style of Example 1 to identify the largest entry in a list. Show how it works on the following lists.

(a)   4, 8, 7, 2, 1   (b)   11, 31, 22, 17, 31

Exercise 3.2   Work through the following algorithm for the numbers 4, 7, 17, 32, 1, 6.

Step 1:   **Input N**
Step 2:   **For X = 1 to N**
Step 3:   **A = S(X)**
Step 4:   **IF A = 2 THEN OUTPUT S(X)**
Step 5:   **IF A < 2 NEXT X**
Step 6:   **A =A ÷2**
Step 7:   **GO TO STEP 4**
Step 8:   **STOP**

Which numbers are outputed at Step 4? What is the algorithm designed to do?

Exercise 3.3   Draw a flow diagram for an algorithm to find the first place where the digit 7 occurs in a list of numbers.

Exercise 3.4   Hero's formula for calculating the area of a triangle when the lengths of the three sides are known, is

$$\text{area} = \sqrt{s(s-a)(s-b)(s-c)}$$

where $s = \dfrac{1}{2}(a+b+c)$ is the semi-perimeter.

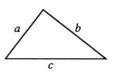

(a) Check that the formula gives the correct answer for the 3, 4, 5 triangle.

(b) Work through the following data set using the given flow diagram.

13, 14, 15, 20, 37, 51, 7, 15, 7, 75, 86, 97

Give the four outputs produced and explain any results you consider unusual.

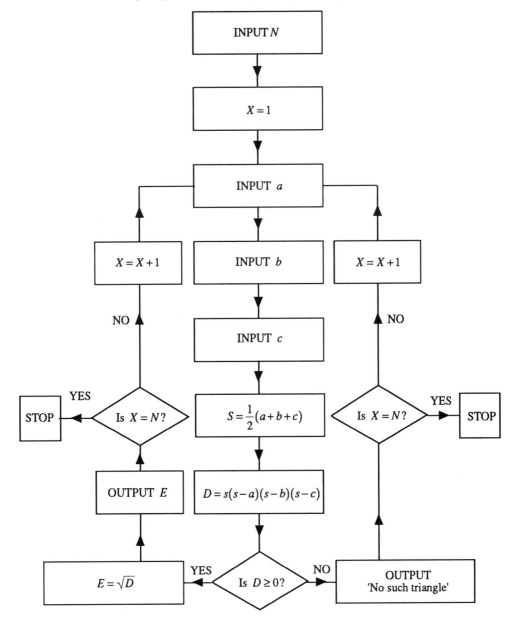

Fig. 16.3

Exercise 3.5   For the following algorithm INT(A) means 'the integral part of A' or 'the largest integer less than or equal to A', so

INT $(7\frac{1}{4})$ = 7

INT (8.9) = 8

INT (3) = 3

INT $(-2\frac{1}{4})$ = -3

Work through the following algorithm showing the final lists produced from the data sets
(a)   9, 8, 4, 3   (b)   9, 8, 11, 2, 3, 7

Explain in words what the algorithm does.

Step 1:   **INPUT N**
Step 2:   **Y = 0**
Step 3:   **Z = 0**
Step 4:   **FOR X = 1 TO N**
Step 5:   **IF Z + Y = N - 2 GOTO STEP 14**
Step 6:   **A = S(X) ÷2**
Step 7:   **IF A = INT(A) GOTO STEP 13**
Step 8:   **B = S(X)**
Step 9:   **S(X) = S(N - Z)**
Step 10:   **S(N - Z) = B**
Step 11:   **Z = Z + 1**
Step 12:   **GOTO STEP 5**
Step 13:   **Y = Y +1**
Step 14:   **STOP**

Exercise 3.6   Fleury's algorithm may be used to decide whether or not it is possible to find a route around a network using each edge exactly once, starting and finishing at the same vertex.

Step 1:   Choose a vertex to start at.

Step 2:   Move along on an edge that has not already been used, disconnecting the network only when it is unavoidable.

Step 3:   Mark the edge used with X.

Step 4:   Repeat STEP 3 and 4 until all the edges have been marked.

Apply the algorithm to the following networks.

In which of these networks does the algorithm provide paths?

(a)   (b)   (c)

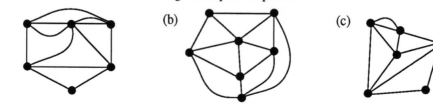

16.4 THE EFFICIENCY OF ALGORITHMS

The basic measure of efficiency is in terms of **time**. The computing speeds of different machines could affect such a concept so the **number of steps** needed to complete the task is used.

Long division has already been met in terms of algorithms; long multiplication can also be written in the form of a step-by-step process. In the case of two n-digit numbers being multiplied, this would involve n^2 multiplications and a possible $n + 1$ carries in the addition part of the calculation.

With two 3-digit numbers you have $3^2 = 9$ multiplications and 4 possible carries, indicated $\uparrow$. The n^2 is larger and so more important than the $n + 1$ term and the algorithm can be said to have order n^2, which may be written $O(n^2)$.

$$
\begin{array}{rrrrr}
 &   & 5 & 6 & 2 \\
 &   & 3 & 1 & 7 \\
\hline
 & 3 & 9 & 3 & 4 \\
 &   & 5 & 6 & 2 \\
1 & 6 & 8 & 6 &   \\
\hline
1 & 7 & 8 & 1 & 5 & 4 \\
 & \uparrow & \uparrow & \uparrow & \uparrow
\end{array}
$$

One approach to an assignment problem would be to use a 'brute force' method, that is, list all the possible assignments that could be made and compare them to find the best.

For example, with five people and five jobs to be done, P1 could be assigned any of the five jobs, P2 any of the four that remain, P3 any of the three that are left, P4 either of the last two and P5 the sole job available. So there are

$$5 \times 4 \times 3 \times 2 \times 1 = 120$$

possible assignments. This can be written 5! for short. In general such a method would have order $n!$

	J1	J2	J3	J4	J5
P1					
P2					
P3					
P4					
P5					

Now consider how $O(n^2)$ and $O(n!)$ algorithms' operating times vary as n increases.

n	1	2	3	4	5	6	10	50
n^2	1	4	9	16	25	36	100	2500
$n!$	1	2	6	24	120	720	3 628 800	3×10^{64}

It is not hard to see why the first algorithm would be classified as efficient while the second certainly would not. An algorithm will be described as efficient if its order is a power of n, in which case it is said to run in **polynomial time**. An inefficient algorithm is one whose number of steps, and therefore running time, increases drastically for larger values of n.

If each step were to take one millionth of a second, i.e. 10^{-6} seconds, then the table would become

n	1	5	10	50
n^2	10^{-6} secs	2.5×10^{-5} secs	10^{-4} secs	2.5×10^{-3} secs
$n!$	10^{-6} secs	1.2×10^{-4} secs	3.6 secs	9.6×10^{5} years

Exercise 4.1   In a two-dimensional simplex method problem with n constraints, as well as the trivial $x \geq 0$ and $y \geq 0$, show that the feasible region has at most $n + 2$ vertices. Why is the simplex method of order n in two dimensions? What is its order in three dimensions?

Exercise 4.2 The tower of Hanoi requires 2^{n-1} moves, so any algorithm to carry out these $\frac{1}{2} \times 2^n$ moves is of order 2^n. Copy and complete the table below and classify such an algorithm as efficient or inefficient. Assume each step takes 10^{-6} seconds.

Steps n	1	5	10	20	50	100
2^n						

Exercise 4.3 Classify the following as efficient or inefficient.

Algorithm	Order
A	n^3
B	3^n
C	n^n

Exercise 4.4 To see if a number is divisible by 7, find the difference between twice the units digit and the number consisting of the remaining digits. If this is divisible by 7, the original number will be also. This process can be written as an algorithm. What is its order?

16.5 USE OF RECURRENCE RELATIONS

The bubble sort algorithm moves the largest element in the list to the end. So

4   7   8   6   1   5   2   3

becomes

4   7   6   1   5   2   3   8

and there are now seven elements to order instead of eight. Moving the largest element involved $n - 1 = 7$ comparisons, 4 with 7, 7 with 8, etc. If the amount of steps with eight is c_8, then

$$c_8 = c_7 + 7$$

and in general

$$c_n = c_{n-1} + n - 1$$

Section 12.6 tells you that c_n is $O(n^2)$ so the bubble sort is efficient.

To find a minimum spanning tree on $t + 1$ vertices from a network with $N + t$ edges it is necessary to remove N edges in total.

How much work is required to move from n superfluous edges to $n - 1$? There are $n + t$ edges in the network to inspect. These can be inspected as in the bubble sort, with the proviso that if an edge's removal disconnects the network, then it is not considered.

So $c_n = (n + t) + c_{n-1}$. Again c_n is $O(n^2)$.

When a shuttle sort has n more numbers to place in position

$N - n$	n

how much work is involved? To put the next one in the correct place takes at most $N - n$ comparisons in the 'worst-case scenario'

$$w_n = N - n + w_{n-1}$$

where w_n is the work needed for n.

Again this leads to $O(n^2)$ for the algorithm.

16.6 NOTES

The following algorithms can be found in this text in the sections listed.

Section	Algorithm	Order
1.3	Kruskal	$O(n^2)$
1.3	Prim	$O(n^2)$
1.4	Chinese postman	$O(n^3)$
1.6	Fleury	$O(n^2)$
2.3	Divisibility	$O(n)$
2.4	Euclid	$O(n)$
3.2	Dijkstra	$O(n^2)$
3.6	Floyd	$O(n^3)$
4.2	Dynamic programming	$O(n^2)$
5.8	Flow augmentation	$O(n^2)$
6.3	Critical path analysis	$O(n)$
8.2	Simplex method	$O(n)$
9.2	Stepping stones method	$O(n)$
10.3	Matching improvement	$O(n)$
10.5	Hungarian algorithm	$O(n)$
14.9	Hero's method	–
15.3	Bubble sort	$O(n^2)$
15.4	Shuttle sort	$O(n^2)$
15.5	Shell sort	$O(n^2)$
15.6	Quick sort	$O(n^2)$

Glossary

Algorithm (or algorism) a step-by-step procedure or set of rules for solving a problem. The word comes from the name of a 9th century Persian mathematician.

Appendix supplementary material at the end of a work.

Assignment problems problems derived from the assignment of people to tasks to optimise a function, e.g. minimise the total time required or maximise the profit involved.

Cardano Italian mathematician of the 16th century.

Cubic a polynomial containing a variable raised to the power three and none to a higher power.

Coefficient the numerical or constant multiplier of a variable in an expression, e.g. the coefficient of x^2 in $4x^2$ is 4.

Critical path the chain of activities which together determine the total time required to complete the project.

Critical path analysis a method to achieve the optimisation of a project by studying the logical relationships between the activities which go to make up the project.

Degeneracy the case occurring when a basic variable has the value zero.

Dijkstra's algorithm a method for finding the shortest route from one node in a network to another by working through nodes in order of increasing distance from the start.

Distinct different.

Dynamic programming the method by solving n-step optimisation problems by reducing the number of steps to $n-1$.

Earliest start time in critical path analysis, the earliest time that a particular activity can start.

Euclid Greek mathematician of the third century BC.

Game theory the theory concerning the best choices of strategy, or strategies, where the idea of competition is involved.

Hero (or Heron) 1st century Greek mathematician.

Homogeneous formed of parts that are all of the same kind; an expression is homogeneous if all its terms have the same degree.
e.g. $y = 5x$ is homogeneous of degree 1,
$x^2y + 5x^3 = 7y^3$ is homogeneous of degree 3.

Hungarian method Kuhn's method for solving assignment problems.

Integer zero or a whole number, positive or negative.

Iteration the repeated application of a function or process to values previously generated by it to produce a sequence (Latin: *itero* repeat).

Latest start time	in critical path analysis, the latest time an activity can start without delaying the completion of a project.
Linear equation	a first degree polynomial equation in one variable, e.g. $y = mx + c$ (graphically: a straight line)
Linear programming	the study of optimising linear functions of non-negative variables subject to linear constraints.
Matching	the pairing of nodes in a network such that those linked are from distinct subsets of nodes within the network.
Minimax strategy	the strategy in game theory based on minimising the maximum loss that can occur, i.e. a play-safe strategy designed to avoid large losses.
Network	a system of arcs and nodes such that every arc has a node at each end and at every point of division.
Perturbation	a slight change in the values of variables.
Planar	a network is planar if it can be drawn on a plane so that arcs only meet at nodes.
Polynomial	a polynomial in a single variable consists of a sum of terms each of which is a product of a constant and the variable raised to a non-negative integral power, e.g. $4x^2 + 0.3x^3 + {}^-8$.
Quartic	a polynomial containing a variable raised to the power four and none higher.
Quintic	a polynomial containing a variable raised to the power five and none higher.
Recurrence relation	an equation that gives a definition for a sequence of values by specifying each one in terms of previously calculated ones, e.g. $r_{n+1} = 2r_n$ $\qquad r_0 = 1$
Recursion	the act of going back in, or reducing, a problem to a smaller one.
Root	a value that solves an equation.
Simplex method	the method of solving a linear programming problem by pivoting to proceed round the vertices of the feasible polytope (n-dimensional version of polygon) in order of increasing (or decreasing) values of the objective form.
Simulation	the formation of a mathematical model to imitate a situation.
Sorting	the arrangement of data into a prescribed order.
Stepping- stones method	Hitchcocks' method for solving transportation problems.
Transportation problems	the class of problems derived from matching supplies from several sources with demands by various consumers so as to minimise cost.
Zero-sum game	a game in which the total payout to those playing is zero, i.e. what one gains, another loses.

Answers

CHAPTER 1 AN INTRODUCTION TO NETWORKS

1.2 Investigations

1. (a), (b), (d), (e) can be drawn. (c), (f) cannot be drawn as they contain an odd number of odd nodes and hence an odd number of ends of lines. This is impossible as each line has two ends and so the total number of ends is twice the number of lines which is even.

2. (a), (b), (d), (f) are traversable and contain 0 or 2 odd nodes. (c), (e) are not traversable as they have more than two odd nodes. Each time a node is passed through in drawing the network two is added to its order so the only ones that can be odd are those where the pen started and finished.

3. (a) (i) 3

    (ii) 6

    (b) $p - 1$

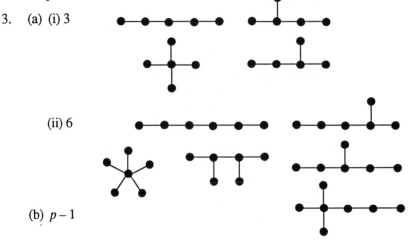

Exercise 2.1   No, the network has four odd nodes and
                 so is not traversable.

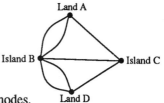

Exercise 2.2   (a) (c) as they have an odd number of odd nodes.
                 (h) (i) (j) (l) as they have an odd number of odd nodes.

                 (b) butane (d) ammonia (e) carbon monoxide (f) carbon dioxide (g) water
                 (k) glucose

Exercise 3.1   (a) 213 miles   (b) 192 miles

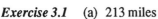

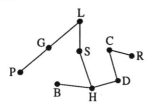

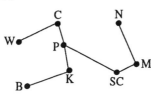

                 (c) 242 miles

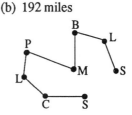

Exercise 4.1   Repeat AC via DG giving an extra 89 km. Total length = 708 km

Exercise 4.2   57.7 cm

Exercise 4.3   74 miles by repeating AE, CD

Exercise 5.1   York 580 km - Manchester 542 km; Leeds 561 km - Sheffield 597 km
                 Newcastle 627 km - Cambridge 719 km
                 Choose 719 km as lower bound. Not optimal.

Exercise 5.2 $2(19 + 22 + 27 + 28 + 34 + 36 + 41) = 414$ miles

Exercise 5.3   309 miles $= 23 + 31 + 62 + 26 + 31 + 44 + 33 + 36 + 23$
                 No. (All sites give the same route as each town is both entered and left.)

Exercise 5.4   296 miles $= 23 + 44 + 17 + 19 + 43 + 37 + 44 + 37 + 32$

Exercise 5.5   (a) 174 miles (Chinese postman)   (b) 76 miles (travelling salesman)
                 (c) Yes 78 miles

Exercise 5.6   AEDCBA = 52 mins

                 AEBCDA = 52 mins

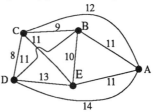

CHAPTER 2   RECURSION

2.2 Investigation

$$u_n = \frac{3}{4} + \frac{1}{4}\left(-\frac{1}{3}\right)^n \qquad u_{100} \approx \frac{3}{4}$$

$$u_0 = 1, \quad u_1 = \frac{2}{3}, \quad u_2 = \frac{7}{9}, \quad u_3 = \frac{20}{27}, \quad u_4 = \frac{61}{81}$$

$$u_n = \frac{4}{5} + \frac{1}{5}\left(-\frac{1}{4}\right)^n \qquad u_{100} \approx \frac{4}{5}$$

See Chapter 12, *Recurrence relations*, for further examples.

Exercise 4.1   (a) yes   (b) no

Exercise 4.2   (a) yes   (b) no

Exercise 4.3   Yes, 456231^2

Exercise 4.4   SADT = 20 miles

Exercise 4.6   21

CHAPTER 3   SHORTEST ROUTE

3.1 Investigations

1.   163 miles via Newark, Nottingham, Derby, Stoke, Crewe

2.   37 hours   S B E T

Exercise 2.1   (a) S A D T 36   (b)   S E G T 32

Exercise 2.2   141 miles via Middlesborough, Newcastle
                 138 miles via Middlesborough, Scotch Corner, Penrith

Exercise 2.3   163 miles via Newark, Nottingham, Derby, Stoke, Crewe

Exercise 5.1   (a) 17 minutes   SAA'CGT or SAA'DGT or SBB'DGT
                 (b) Not affected as there is a route SAA'CGT not via D. (If this had not existed
                     it would have been quickest to use Dijkstra's algorithm on the network
                     with D and incident arcs deleted to seek a route quicker than 20 minutes.)

Exercise 5.2   E:20   F:15   H:18
                 CE + FH = 23   CF + EH = 21   CH + EF = 23
                 so repeat CF and EH (CF is via G)
                 e.g. ABDEHEFHDGBCGFGCA
                 length = 96

Exercise 5.3   (a) 80 mins PADECBQ   (b) 84 mins PDECBQ

Exercise 5.4   34 mins   SDCBT

CHAPTER 4 DYNAMIC PROGRAMMING

4.1 Investigations

1.   X B D Y = 41

2.   S A E L T = 25

Exercise 2.1   (a) S B F K T = 24   (b) S C H L T = 32

Exercise 2.2   (a) SBDFHIT = 22   (b) SBCFGIT = 29 or SBCFGJT = 29

Exercise 2.3   A, B2, C2, D1, E = 11 or A, B2, C2, D4, E = 11

Exercise 3.1

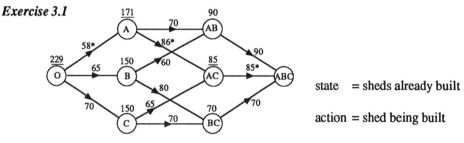

state   = sheds already built

action = shed being built

maximum profit = £229 by the schedule A, C, B

Stage	State	Action	Return	Value
1	AB		90	90
	AC		85	85
	BC		70	70
2	A	B	70	90 + 70 = 160
		C	86	85 + 86 = 171 *
	B	A	60	90 + 60 = 150 *
		C	80	70 + 80 = 150 *
	C	A	65	85 + 65 = 150 *
		B	70	70 + 70 = 140
3	0	A	58	171 + 58 = 229 *
		B	65	150 + 65 = 215
		C	70	150 + 70 = 220

maximum profit = £229   schedule = ACB

Exercise 3.2

$P \rightarrow S \rightarrow R \rightarrow Q$
or
$R \rightarrow Q \rightarrow P \rightarrow S$
each gives 25 hours

Stage	State	Action	Value
1	PQR	S	$\underline{7}$
	PQS	R	1
	PRS	Q	$\underline{5}$
	QRS	P	2
2	PQ	R	$3+7 = 10*$
		S	$10+1 = 11$
	PR	Q	$9+7 = 16*$
		S	$11+5 = 16*$
	PS	Q	$9+1 = 10$
		R	$2+5 = \underline{7}*$
	QR	P	$4+\underline{7} = \mathbf{11}*$
		S	$12+2 = 14$
	QS	P	$3+1 = 4*$
		R	$2+2 = 4*$
	RS	P	$3+5 = 8*$
		Q	$7+2 = 9$
3	P	Q	$11+10 = 21$
		R	$4+16 = 20$
		S	$12+\underline{7} = \underline{19}*$
	Q	P	$5+10 = 15*$
		R	$4+11 = 15*$
		S	$13+4 = 17$
	R	P	$5+16 = 21$
		Q	$9+\mathbf{11} = \mathbf{20}*$
		S	$14+8 = 22$
	S	P	$4+7 = 11$
		Q	$10+4 = 14$
		R	$3+8 = 11*$
4	0	P	$6+\underline{19} = \underline{25}*$
		Q	$12+15 = 27$
		R	$5+\mathbf{20} = \mathbf{25}*$
		S	$15+11 = 26$

PSRQ or RQPS $\Rightarrow$ 25 hours

Exercise 3.3

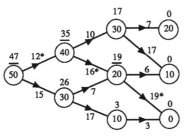

Stage	State	Action	State	Return
1	30	A		17*
		D		7
	20	A		$\underline{19}$ *
		D		6
	10	D		3*
2	40	A	20	$16+19=\underline{35}$ *
		D	30	$10+17=27$
	30	A	10	$17+3=20$
		D	20	$7+19=26$ *
3	50	A	30	$15+26=41$
		D	40	$12+35=\underline{47}$ *

$\Rightarrow$ best is defend, attack, attack $\Rightarrow$ value = 47

Exercise 3.4

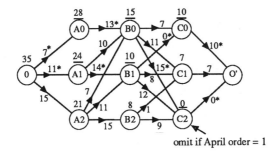

omit if April order = 1

Schedule : 1, 3, 3, 0 or 2, 3, 0, 2   Minimum cost = £3500

Stage	State	Action	Cost
1	C0	2	$\underline{10}$
	C1	1	7
	C2	0	$\underline{0}$
2	B0	1	$7+10=17$
		2	$11+7=18$
		3	$15+\underline{0}=\underline{15}$ *

cont'd

Stage	State	Action	Cost
	B1	0	$0 + \underline{10} = \underline{10} *$
		1	$8 + 7 = 15$
		2	$12 + 0 = 12$
	B2	0	$1 + 7 = 8*$
		1	$9 + 0 = 9$
3	A0	3	$13 + \underline{15} = \underline{28} *$
	A1	2	$10 + 15 = 25$
		3	$14 + \underline{10} = \underline{24} *$
	A2	1	$7 + 15 = 22$
		2	$11 + 10 = 21*$
		3	$15 + 8 = 23$
4	0	1	$7 + \underline{28} = \underline{35} *$
		2	$11 + \underline{24} = \underline{35}$
		3	$15 + 21 = 36$

CHAPTER 5   FLOWS IN NETWORKS

Investigation 1   Max flow 155 cars

Investigation 2   Max flow 33 cars

Exercise 7.1

(a)   max flow 18   cutset (SB, AD, CT)

(b)   max flow 16   cutset (IT, HT, JT)

(c)   max flow 11   cutset (SK, LK, LM, LO)

(d)   max flow 19   cutset (SR, QR, QV, QU, QP, SP)

(e)   max flow 12   cutset (SB, BC', C'C", AD)

(f)   max flow 21   cutset (HK, FK, I'I", GJ)

Exercise 7.2

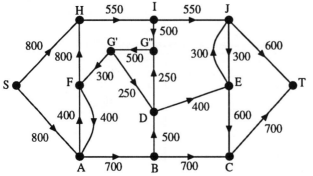

max flow = 1250 cars/hour

cutset (AB, FG', HI)

Exercise 8.1

(a) 6   (b) 8   (c) 50

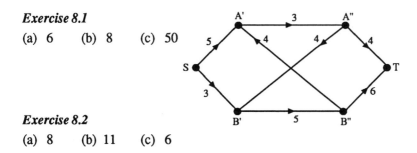

Exercise 8.2

(a) 8   (b) 11   (c) 6

Exercise 9.1

(a) feasible   (b) not feasible   (c) feasible

Exercise 9.2

(a) 11 (SA, AB, BT)   (b) 8 less than sum of lower capacities on CT + DT, (BC, DC, CT)

(c) 10 (AC, CB, BD)

Exercise 9.3   max flow = 14 (SAT-4, SABT-2, SBT-8)

Exercise 9.4   max flow = 14 (SBDT-4, SADT-4, SACT-6)   cutset = (CT, DT)

Exercise 9.5 $b \geq 2$, $a \leq 3$

CHAPTER 6   CRITICAL PATH ANALYSIS

Investigation 1   (a) 60 mins as laid out   (b) 2

(c) If the person not mixing the batter prepares the dessert, 5 minutes are saved.

(d) Yes. The dessert chills while the first course is eaten.

Exercise 3.1

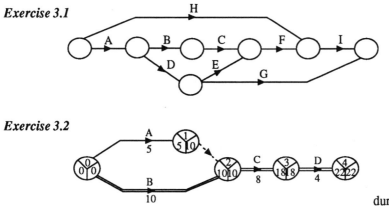

Exercise 3.2

duration = 22 days

Note   The identity dummy can be placed before or after either A or B. The delay
increases the total time by one day.

Exercise 3.3

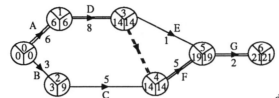

duration = 21 days

critical path: A D F G   delays: B 6, C 6, E 4   (B + C = 6)

Note For E, EST = LST at head and tail but it is **not** critical.

Exercise 3.4

critical path: A D G H K.   duration = 24 hours

Exercise 3.5

duration = 15 hours

critical path: B' C F G K. The identity dummy may be placed before or after H or I.

Exercise 3.6

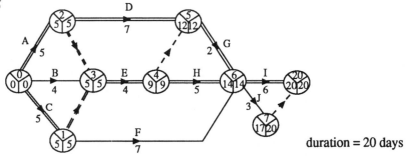

duration = 20 days

critical paths: A D G I, C E H I, A E H I
non critical activities: B F J (do not appear on critical paths)
(The identity dummy 7 - 8 has other possible positions as it may be placed before or after I
or J.)

Exercise 6.1

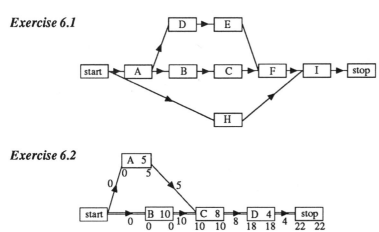

Exercise 6.2

duration = 22 days

The delay increases the total time by one day.

Exercise 6.3

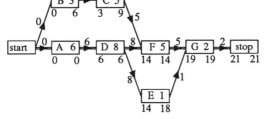

duration = 21 days

critical path: A D F G   delays: B 6, C 6, E 4 (C + B = 6)

Exercise 6.4

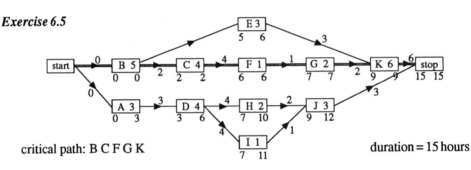

(d) arc A D is re-labelled with 2.

Exercise 6.5

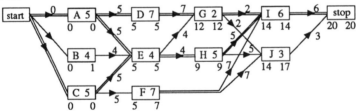

critical path: B C F G K   duration = 15 hours

Exercise 6.6

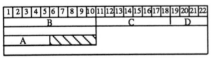

critical paths: A D G I, A E H I, C E H I   non-critical activities: B, F, J.

Exercise 9.1   *Exercise 9.2*

Exercise 9.3

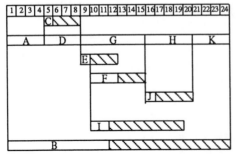

J is dependent on I.
As I slides along its bar, the fence
which starts at 11 will 'catch' J if I starts
at 13 or later.

Exercise 9.4

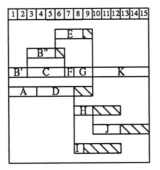

Exercise 10.1
(a) No change profitable.   (b)

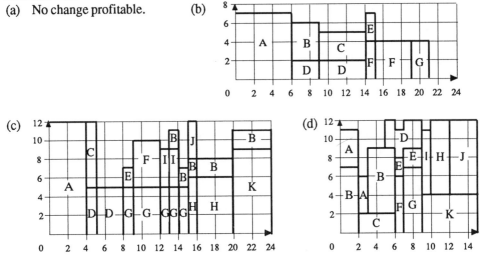

(c)

(d)

CHAPTER 7 LINEAR PROGRAMMING (GRAPHICAL)

Investigation 1

(a) 110, all without arms   (b) 40   (c) 100 all without arms at a cost of £2000
(d) 104: 21 with arms and 83 without; 40 (no change); 20 with arms, 80 without £2100

Exercise 2.1   5 eight-seaters + 8 five-seaters $\Rightarrow$ £220
Yes, 5 eight-seaters + 9 five-seaters or 4 eight-seaters + 10 five-seaters.

Exercise 2.2   (a) 5 (b) 6   Cheapest: 2 overhauls + 6 services = £400

Exercise 2.3   84 trees + 120 shrubs

Exercise 2.4 $20x + 24y$, $40x + 16y$
$3x + 2y \geq 60$, $2y \leq 5x$ Cheapest: 8 set A + 18 set B = £64

Exercise 3.1   8 bats + 24 rackets $\Rightarrow$ £200

Exercise 3.2 $5\frac{3}{8}$ h of carrots + $3\frac{3}{8}$ h of potatoes $\Rightarrow$ £2017.50

Exercise 3.3   100 ft of A and 150 ft of B $\Rightarrow$ 250 ft

Exercise 3.4 $\dfrac{x}{4} + \dfrac{2y}{3} \le 600 \Rightarrow 3x + 8y \le 7200$

$3x + y \le 3000$

$(x, y) = (800, 600) \Rightarrow 1400$ metres

CHAPTER 8   LINEAR PROGRAMMING: SIMPLEX METHOD
Investigation

$0.4x + 0.35y + z \le 69.9$

$\Rightarrow \quad 8x + 7y + 20z \le 1398$

$20x + 15y + 10z \le 2700$

$\Rightarrow \quad 4x + 3y + 2z \le 540$

$$y \le \frac{1}{3}(x + y)$$

$\Rightarrow -x + 2y \le 0$

$P = x + y + z$

Following the graphical chapter a logical approach would be to find the points of intersection of the six planes, $8x + 7y + 20z = 1398$, $4x + 3y + 2z = 540$, $-x + 2y = 0$, $x = 0$, $y = 0$, $z = 0$, which have positive coordinates. Solution is

$x =$ no. of chairs with arms $= 92$, $y =$ no. basic chairs $= 46$, $z =$ no. of tables $= 17$

Exercise 3.1 $P = 1400$ $x = 800$ $y = 600$

Exercise 3.2 $P = 150$ when $x = 75$, $y = 0$

Exercise 3.3 $P = 50$ when $x = 20$, $y = 30$

Exercise 3.4 $P = 700$ $x = 300$ $y = 200$

Exercise 3.5 $P = 27.4$ $x = 5.6$ $y = 1$

Exercise 3.6   8 bats + 24 rackets $\Rightarrow$ £200

Exercise 3.7 $5\dfrac{3}{8}$h of carrots + $3\dfrac{3}{8}$ h of potatoes $\Rightarrow$ £2017.50

Exercise 6.1 $P = 10$ when $x = 2$, $y = 4$

Exercise 6.2 $P = 54$ when $x = 0$, $y = 1$, $z = 5$

Exercise 6.3 $P = 20$ when $x = 20$, $y = 10$

Exercise 6.4 $250x + 175y + 75z = P$

$5x + 3y + z \le 65$

$x + y + z \le 18$

$16x + 9y + 5z \le 205$

$P = £3562.50$ when $x = 5\dfrac{1}{2}, y = 12\dfrac{1}{2}, z = 0$

Exercise 6.5   1600 m = 200 m walking + 1400 m trotting

Exercise 6.6   Total $= 148\frac{4}{7}$ when $x = 42\frac{6}{7}$, $y = 51\frac{3}{7}$, $z = 54\frac{2}{7}$

CHAPTER 9 THE TRANSPORTATION PROBLEM

Exercise 2.1

2		2
2	3	

Exercise 2.2

6	14	
1		18
16		

Exercise 2.3

	15	5
12		
5		10
8		

Exercise 2.4

	200		300
300			300
	200	500	

Exercise 6.1, e.g.

Not unique
Could utilise *
Total = 2637

	33	
6	30	3
	21	*
51		
*		72

Exercise 6.2

	1	2
1	1	
		1

1 boat left at D2

cost = £71

Exercise 6.3

10		15		15
2	8			
	0		20	
	10			

cost = £428 000

Exercise 6.4

50		
50	50	
		60

or

30		
	50	
20		60

10 loads left at Q1   30 loads left at Q1

20 loads left at Q3

CHAPTER 10 MATCHING AND ASSIGNMENT PROBLEMS

Investigation 1   e.g. W1 J1, W2 J5, W3 J3, W4 J4, W5 J2

Investigation 2   No, there is no possible pairing.

Exercise 3.1

(a) A2 B3 C5 D4 E1 or A1 B3 C2 D4 E5

(b) No matching is possible for the complete sets as A, C, E only connect to 2, 3.

(c) A3, B2, C1, D5, E4, F6 or A3, B2, C5, D1, E4, F6

(d) A1, B4, C2, D3, E5 or A2, B4, C3, D1, E5 or A1, B5, C2, D4, E3 and note
     also A4, B5, C2, D1, E3

Exercise 3.2   AX   BY   CW   DZ
                 AZ   BY   CW   DX
                 AX   BW   CZ   DY
                 AY   BW   CZ   DX

Exercise 3.3   No staffing is possible to cover all six classes with 'expert' teachers.

Exercise 3.4   A1, B3, C2, D4 ⎫
                 A2, B3, C1, D4 ⎬   3 ways
                 A4, B2, C1, D3 ⎭
                 A4, B2, C1, D3 ⇒ total = 47.7

Exercise 3.5   Several answers, e.g. S47, C23, B16, T58

Investigation 3
Leif - 10 km, Mark - cross-country, Nathan - 5 km; gives a minimum total of
$20 + 9 + 12 = 41$ minutes.

Investigation 4
Firm 1, A; Firm 2, B; Firm 3, C; Firm 4, D for a minimum cost of
$100 + 160 + 170 + 100 = £530$

Exercise 5.1   Answer as given in the example but arrived at more quickly.

Exercise 5.2   As for Investigation 4.

Exercise 5.3   Mancastle-Sheffingham, Newbridge-Oxfield, Camdon-Brisford,
Lanchester-Nottol; distance $= 97 + 47 + 104 + 83 = 331$ miles, or 662 miles for lorries to
return.

Exercise 5.4   Ashoke-packer, Fiona-secretary/receptionist, Gemma-computer operator,
Karl-accounts clerk, Tamsin-driver. Minimum cost $= 85 + 100 + 100 + 140 + 95 = £520$.

Exercise 7.1   Maximum value $= 34$ by using the assignment P1 J1, P2 J3, P3 J2.

Exercise 7.2   Minimum value $= 40$ by using the assignment P1 J1, P2 J3, P3 J2, P4 J4.

Exercise 7.3   Ann-hurdles, Carol-100 m, Denise-200 m, Thani-long jump, Wendy-high
jump, as total averages $= 18.6$.

Exercise 9.1   Long-Mechanics, Evans-Pure, Newton-Discrete, Francis-Statistics
(Philips-none); minimum total $=13.6$.

Exercise 9.2   Aston-Adams, Sheffield-Pinder, Chelocean-Patel, East Bromwich-Cooper,
West Spam-Jones; maximum $= £205,000$

Exercise 9.3   Naomi-3, Sarah-5, Ruth-1, and either Deborah-2 and newcomer -4
or Deborah-4 and newcomer-2. Minimum time for the four operators $= 136$ mins.

Exercise 9.4   A1-D2, A2-D3, A3-D1, A4-D4; total estimated casualties $= 440$.

CHAPTER 11   GAME THEORY

Investigation

p_1 should be adopted $\dfrac{7}{15}$ of the time, p_2 $\dfrac{8}{15}$ of the time and p_3 not at all if the aim is to minimise losses. A number of other combinations are possible if the reason for choosing tactics is not to avoid large losses but, for example, to try for big gains.

Exercise 2.1   (a), (b), (c), (e) are stable, (d) is unstable

Exercise 2.2   A stable solution is the smallest entry in its row and the largest in its column.

Exercise 5.1   (a) $\begin{array}{|cc} 3 & 5 \\ 6 & 3 \end{array}$ (b) $\begin{array}{|cc} 2 & 1 \\ -1 & 2 \end{array}$ (c) $\begin{array}{|c} 1 \end{array}$ (d) does not reduce

Exercise 5.2   (a) no stable solution   (b value $= -1$; a1 b3 or a4 b3
(c) value $= 4$; a2 b2, a2 b4, a4 b2, a4 b4   (d) no stable solution

Exercise 8.1

(a)   A: $P(a1) = \dfrac{5}{8}$ $P(a2) = \dfrac{3}{8}$ value $= \dfrac{11}{8}$

       B: $P(b1) = \dfrac{7}{8}$ $P(b2) = \dfrac{1}{8}$ value $= -\dfrac{11}{8}$

(b)   A: $P(a1) = \dfrac{2}{5}$ $P(a2) = \dfrac{3}{5}$ value $= \dfrac{14}{5}$

       B: $P(b1) = \dfrac{2}{5}$ $P(b2) = \dfrac{3}{5}$ value $= -\dfrac{14}{5}$

Exercise 8.2 (Investigation)

       M: $P(m1) = \dfrac{1}{3}$ $P(m2) = \dfrac{2}{3}$ value $= -\dfrac{2}{3}$

       P: $P(p1) = \dfrac{7}{15}$ $P(p2) = \dfrac{8}{15}$

                            $P(p3) = 0$ value $= \dfrac{2}{3}$

Exercise 9.1

(a)   A: $P(a1) = \dfrac{2}{5}$ $P(a2) = \dfrac{3}{5}$ value $= \dfrac{14}{5}$

(b)   A: $P(a1) = \dfrac{5}{8}$ $P(a2) = \dfrac{3}{8}$ value $= \dfrac{11}{8}$

Exercise 9.2

(a) $P(a1) = \dfrac{6}{7}$ $P(a2) = \dfrac{1}{7}$ $P(a3) = 0$ value $= \dfrac{103}{7}$

Exercise 10.1   (a) not stable   (b) a1b2 $= 3$ is stable
(c) a1b2 $=$ a1 b3 $=$ a2 b2 $=$ a2 b3 $= 2$ is stable

Exercise 10.2 (a)

$$\begin{array}{ccc} 0 & 14 & 2 \\ 8 & 6 & 7 \end{array}$$

(b) make B's gains positive, i.e. entries negative. Subtract at least 14.

$$\begin{array}{ccc} -14 & 0 & -12 \\ -6 & -8 & -7 \end{array}$$

$E(\text{B plays a1}) = 14p_1 + 12p_3$

$E(\text{B plays a2}) = 6p_1 + 8p_2 + 7p_3$

$$\begin{array}{ll} v - 14p_1 - 12p_3 & \leq 0 \\ v - 6p_1 - 8p_2 - 7p_3 & \leq 0 \\ p_1 + p_2 + p_3 & \leq 0 \end{array}$$

Exercise 10.3

	b1	b2
a1	2	1
a2	-2	7

minimax $= 2 \neq 1 =$ maximin

A: $P(\text{a1}) = 0.9$ $P(\text{a2}) = 0.1$ value $= 1.6$

B: $P(\text{b1}) = 0.6$ $P(\text{b2}) = 0.4$ value $= -1.6$

CHAPTER 12   RECURRENCE RELATIONS

Investigation (a) $u_n = 4 \times 2^n - 3 = 2^{n+2} - 3$ $u_{100} = 2^{102} - 3 \approx 5.1 \times 10^{30}$

Investigation (b) $u_n = 1$ $u_{100} = 1$

Investigation (c) $u_n = 2 \times 0.5^n + 2 = 0.5^{n-1} + 2$ $u_{100} = \dfrac{2}{2^{100}} + 2 \approx 2$

Investigation (d)   possible by this method, but much harder

$$u_n = 2 \times 2^n - n - 1 = 2^{n+1} - n - 1$$

$$u_{100} = 2 \times 2^{100} - 101 \approx 2.5 \times 10^{30}$$

Exercise 3.1   (a) $7^n - 1$ (b) $3.5 \times 5^n - 0.5$

(c) 10×0.8^n (d) $3.5 - 2.5(-1)^n$

Exercise 3.2 $x_n = 1350(1.08)^n - 1250$. $x_{19} = £4576.20$ (to the nearest penny)

Exercise 3.3 m_3 to put the top three on post B, + 1 to put the largest disc on C,

+ m_3 to move the small three from B to C

$$m_{n+1} = 2m_n + 1$$

$$m_n = 2^n - 1$$

Exercise 4.1   (a) $A\left(\dfrac{1}{3}\right)^n + \dfrac{3}{2}$ (b) $A(5)^n + 1$ (c) $A \times 7^n - 2n - \dfrac{1}{3}$

Exercise 4.2 $2^n - n - 1$

Exercise 6.1   (a) $\dfrac{2}{3} \times 2^n + \dfrac{1}{3} \times 5^n$ (b) $0.9 \times (-3)^n + 0.1 \times 7^n$

(c) $2 \times 3^n + n \times 3^{n-1}$ (d) n^2 (e) $2n$

Exercise 6.2 $u_{n+1} = u_n + n - 1 \Rightarrow u_n = \frac{1}{2}n^2 - \frac{3}{2}n$

Exercise 6.3. $\frac{1}{4} \times 5^n + \frac{3}{4}$

Exercise 7.1   (a) $3(-2)^n + 5 \times 4^n$ (b) $8(-0.5)^n + 3^{n+1}$

(c) $3^n + 2n \times 3^n$ (d) $3 \times (0.5)^n - n(0.5)^n$

Exercise 7.2 $c_n = c_{n-1} + 6c_{n-2}$

$c_n = 5 \times 3^n + 3(-2)^n$

71 646 231

Exercise 7.3 $2^{n+1} + (-2)^n$ *Exercise 7.4* $5\left(\sqrt{3}\right)^n - 4\left(-\sqrt{3}\right)^n$

Exercise 9.1 $2 \times 3^n - 3 \times 2^n + 1$ *Exercise 9.2* $4.5(2)^n + 0.5(-2)^n - 3$

Exercise 10.1 3×7^n *Exercise 10.2* $5n$

Exercise 10.3 $2^{n+1} - 3$ *Exercise 10.4* $\frac{5}{3}(0.2)^n + \frac{10}{3}(0.5)^n$

Exercise 10.5 $\frac{31}{5}(4)^n - \frac{1}{5}$ *Exercise 10.6* $9^n - 2(4)^n$

Exercise 10.7 $(5+n)\left(-\frac{1}{3}\right)^n$ *Exercise 10.8* $C = 1, D = 4$

Exercise 10.9 $A2^n + B3^n + 7$ *Exercise 10.10* $u_n = 5 \times 3^n - 4 \times 2^n$

Exercise 10.11   (a) $\phi_n = 2\phi_{n-1}$ (b) $\theta_n = 3\theta_{n-1} + \phi_{n-1}$ $\phi_n = 2^n, \ \theta_n = 2 \times 3^n - 2^n$

Exercise 10.12 $r_{n+1} = r_n + n + 1; \ \ r_n = \frac{n^2}{2} + \frac{n}{2} + 1; \ \ r_{100} = 5051$

Exercise 10.13 $30\,000 \times 1.12 - 12M; \ \ \lambda = 1.12 \quad M = £318.75;$

$P_n = 31875 - 1875(1.12)^n$

Exercise 10.14 $\omega_{n+1} = \frac{3}{4}\omega_n; \ \ \omega_n = \left(\frac{3}{4}\right)^n; \ \ t_n = 0.2\left(\frac{3}{8}\right)^n + 0.8; \ \ \omega_n \to 0 \qquad t_n \to 0.8$

Exercise 10.15 $u_n = 20c - 4c\left(\frac{4}{5}\right)^n; \ \ u_n \to 20c$

Exercise 10.16 $D_t = 286 - 0.1D_{t-1}; \ \ D_t = A(-0.1)^t + 260; \ 260; \ 9.2\%$

CHAPTER 13   SIMULATION

Exercise 3.1   0-29 rejected on letter   L   0-29 fail first interview   I
                      0-24 fail aptitude test   A   otherwise go for final interview   F

(a) F F L F L L L L I A   (b) 3

(c) $0.7 \times 0.7 \times 0.75 = 0.3675$ (could draw a tree or part of one)

(d) $10 \times 0.3675 = 3.675 \Rightarrow$ 4

Exercise 3.2 $0 - 64 = P, \quad 65 - 99 = F$ length from tally frequency

										length from	tally
F	F	P	F	P	P	P	F	P	F	0	卌 卌 I
F	P	P	F	F	F	P	P	P	P	1	卌
P	P	P	F	P	P	F	P	P	P	2	IIII
P	P	F	P	F	P	P	P	F	P	3	卌 I
P	P	F	F	P	F	P	P	P	P	4	
P	P	P	P	P	P	P	P	P	F	5	II
P	P	P	F	F	P	P	P	F	P	6	
P	P	P	P	P	P	F	P	P	P	7	II
F	F	F	F	P	F	P	P	F	F		
F	P	P	P	P	P	F	P	P	F	13	I

(c) $P(0) = 0.35$

    $P(1) = 0.2275$

    $P(2) = 0.147875$

    $P(5) = 0.0406$ (to 3 s.f.)

    $P(13) = 0.00129$ (to 3 s.f.)

    13 is very unlikely. Its occurrence
    is surprising.

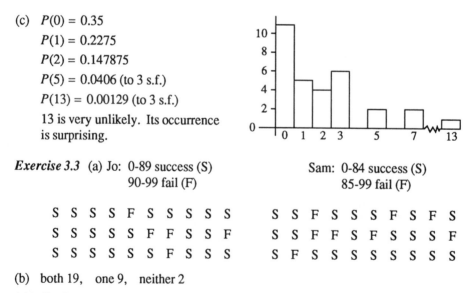

Exercise 3.3 (a) Jo: 0-89 success (S)   Sam: 0-84 success (S)
                          90-99 fail (F)   85-99 fail (F)

S S S S F S S S S S   S S F S S S F S F S
S S S S S F F S S F   S S F F S F S S S F
S S S S S S F S S S   S F S S S S S S S S

(b) both 19,   one 9,   neither 2

(c) (i) $0.1 \times 0.15 = 0.015$ (ii) $0.1 \times 0.85 + 0.9 \times 0.15 = 0.085 + 0.135 = 0.22$

    (iii) $0.9 \times 0.85 = 0.765$

(d) Twice [10, 11, 12, and 28, 29, 30]

(e) $(0.9 \times 0.85)^3 = 0.765^3 = 0.448$ (to 3 s.f.)

(f) Twice in 10 rounds $\Rightarrow 0.2$, very different from 0.448 as only small sample.

Exercise 3.4   For A: 0-59 success (A)   For B: 0-49 success (B)
                     60-99 failure (a)   50-99 failure (b)

            A B A b a B a B a b a B   B wins 1st game 4-2 $\Rightarrow$ 0-1
            b a b A b a b a B A b A   A wins 2nd game 3-1 $\Rightarrow$ 1-1
            a B a B A b a B   B wins 3rd game 3-1 $\Rightarrow$ 1-2
            b A b a b A b A   A wins 4th game 3-0 $\Rightarrow$ 2-2
            A B A b A   A wins 5th game 3-1 $\Rightarrow$ 3-2
            b A b A B a B A b A   A wins 6th game 4-2 $\Rightarrow$ 4-2
            A b a B a b A b A   A wins 7th game 3-1 $\Rightarrow$ 5-2
                          1st set to A : 5-2.

Exercise 3.5   Using single digits : 1 = has coupon (C); 2-8 = no coupon (–); 0,9 discarded.

(a)
     –CC –   – – – –   – – – –   – – – –   – – – –
     – – – –   – – – –   – – – –   – – – –   – – –C
     – – – –   – – – –   C – – –   – – –C   – – – –
     – – – –   – – –C   – – –C   – – – –   CC – –
     – – – –   –C – –   – – – –   –C – –   –C – –

(b)   (i) 15   (ii) 8   (iii) 2   (iv) 0   (v) 0   (vi) 0   (c) $\dfrac{1\times 8 + 2\times 2}{25} = 0.48$

(d) $\dfrac{7}{8}\times\dfrac{7}{8}\times\dfrac{7}{8}\times\dfrac{7}{8} = \dfrac{2401}{4096} = 0.586$ (to 3 d.p.)   (e) $\dfrac{2401}{4096}\times 25 = 14.65\ldots \Rightarrow 15$

CHAPTER 14   ITERATIVE PROCESSES

14.2 Investigation (a)

(i)   1.132997567   (ii) $x_7 = -0.0037$ then further results are impossible.

(iii) 1.133035178

(iv) $x_5 = -3.63\times 10^{36}$ then further values are beyond the range of most calculators.

14.3 Convergent Sequences $x_n = \dfrac{3 - x_{n-1}}{x_{n-1}^4}$ does **not** converge

$x_1 = 1 \Rightarrow x_6 = 7.77\times 10^{63}$ then x_7 is beyond most calculators to compute but is effectively zero so $x_8 = \infty$.

Exercise 3.1 $x_{10} = 7.967574617$, $x = 8$ ***Exercise 3.2***   0.037037, $\dfrac{1}{27}$

Exercise 3.3   2.4251   ***Exercise 3.4***   1.414

Exercise 3.5   (a) $x_n = \dfrac{10}{x_{n-1}} + 3$ (i) 3.283   (ii) yes, both do   (b) –1.32

(c) there at least two different solutions to the equation.

Exercise 3.6 2.206   *Exercise 3.7* 3.181

Exercise 3.8 2.28

Exercise 5.1 $x^3 = 100$, 4.642

Exercise 5.2 (a) 1.414 (c) −1.414

Exercise 5.3 $x_{10} = 1.475931147$, $\dfrac{x_9 + x_{10}}{2}$, 1.467380513

Exercise 5.4 (a) 4.257 (b) 0.4033 (c) −4.66

Exercise 9.1 (a) 6.245 (b) 4.123   *Exercise 9.2* Does not converge

Exercise 9.3 $\sqrt[3]{12}$ *Exercise 9.4* $\sqrt[4]{12}$

Exercise 9.5 $x_n = \dfrac{1}{9}\left(8x_{n-1} + \dfrac{12}{x_{n-1}^{11}}\right)$ *Exercise 9.6* $-1, \dfrac{1}{2}, -\dfrac{1}{2}$

Exercise 9.7 2.186   *Exercise 9.8* −5.57, 3.95

Exercose 9.9 2.78

CHAPTER 15 SORTING AND PACKING

Investigation This can be done in 9 moves e.g. (1, 6) (1, 4) (6, 3) (6, 5) (6, 2) (5, 2) (2, 3) (2, 4) (3, 4)

Bubble sort   9 exchanges. The other two are (4, 2), (3, 2).

Exercise 7.1 and 7.2 (a)

	Bubble	Shuttle	Shell
C	28	28	44
E	28	28	12
T	84	84	68
(b) C	28	28	44
E	7	7	7
T	42	42	58

No sort is universally best. When there are few exchanges to be made the bubble and shuttle sorts are preferable but in most cases the shell sort is likely to be best.

Exercise 7.3   (a)

	Bubble	Quick
C	10	7
E	6	4
(b) C	21	12
E	13	4

Note: Quick sort results are for moving pointer on the left at the start. Quick sort is clearly better.

Exercise 7.4

	Shell	Quick
C	26	13
E	9	7
T	44	27

Exercise 7.5

(a) The bubble sort just shunts the element along in the right direction one place at a time, whereas the shuttle sort scans back down the array for the correct position.

(b) Both sorts are efficient at moving elements forwards to their proper positions, but if one needs to move backwards the bubble sort can only move it one place on each pass.

Exercise 7.6 $8, 4, 2, 1.$

Exercise 7.7   Each sub-set is quick sorted without reference to the rest of the array. So the quick sort is applied to smaller and smaller sub-sets recursively.

Exercise 7.8   The quick sort's worst-case scenario is unlikely to occur. The bubble sort requires a time of the order of n^2 to sort most arrays, but the time required for the quick sort to sort an average array is proportional to $n \log_2 n$, or something similar, i.e. much smaller than n^2 for large n.

Exercise 9.1   Possible. $B + G = 24 + 8 = 32$

$$F + A + E = 22 + 6 + 4 = 32$$
$$I + C + H = 16 + 9 + 7 = 32$$
$$D + J = 17 + 15 = 32$$

Exercise 9.2   4 $200 + 100$

$$175 + 125$$
$$175 + 75 + 50$$
$$150 + 75 + 75$$

Exercise 9.3   Yes. $11 + 5$ $= 16$

$$9 + 4 + 3 \ = 16$$
$$9 + 6 \qquad = 15$$
$$8 + 5 + 3 \ = 16$$
$$6 + 6 + 4 \ = 16$$

but this is not produced by the decreasing first fit algorithm.

Exercise 9.4   No

Exercise 9.5 $9 + 3, \quad 8 + 4, \quad 8 + 4, \quad 7 + 5, \quad 6 + 5 + 1, \quad 5 + 4 + 3$

Exercise 9.6 $12 + 3, \quad 10 + 5, \quad 10 + 5, \quad 9 + 6, \quad 8 + 7, \quad 6 + 4 + 3 + 2$

CHAPTER 16   ALGORITHMS

16.1   The sequence is $1, 3, 7, 15, 31, 63, 127, 255, 511, 1023$ with general term $2^n - 1$.

16.2 Investigations

1. e.g. $B \div A$

STEP 1	Note the first digit of B
STEP 2	Divide by A giving the answer as a whole number and remainder
STEP 3	Write down the number
STEP 4	Note the remainder
STEP 5	If there are no unused digits in B go to step 9
STEP 6	Multiply the remainder by 10 and add the next digit of B
STEP 7	Divide by A giving the answer as a whole number and remainder
STEP 8	Repeat steps 3-7 until all the digits of B have been used
STEP 9	If the remainder is zero go to step 17

STEP 10   Put a decimal point after the last number
STEP 11   Multiply the remainder by 10
STEP 12   Divide by A giving the answer as a whole number and remainder
STEP 13   Write down the number
STEP 14   Note the remainder
STEP 15   If the remainder is zero go to step 17
STEP 16   Repeat steps 11-15 until there are three numbers after the decimal
               point
STEP 17   STOP

2. (a) 25614   (b) 90127

    'Bring down' zeroes two at a time after the decimal point. e.g. $\sqrt{A}$

STEP 1   Split the digits into pairs from the decimal point allowing a single figure
               furthest from the point if necessary
STEP 2   Write down the single figure if it exists and the first pair if it does not
STEP 3   Find the largest digit, X, which when squared, can be subtracted from this
               number
STEP 4   Carry out the subtraction noting the remainder, R
STEP 5   If there are no pairs of digits go to STEP 11
STEP 6   Multiply the remainder by 100 and add the next pair, calling it T
STEP 7   Find the largest digit Y so that $(20X + Y) \times Y \le T$
STEP 8   Work out $T - (20X + Y) \times Y$ and call it R
STEP 9   Replace the old value of X by a new one, $10X + Y$
STEP 10   Repeat steps 6-9 until there are no unused pairs
STEP 11   If R is zero go to STEP 17
STEP 12   Multiply R by 100 and call it T
STEP 13   Find the largest digit Y so that $(20X + Y) \times Y \le T$
STEP 14   Replace the value of X by $10X + Y$
STEP 15   Repeat steps 12-14 three times
STEP 16   STOP: $\sqrt{A}$ is $\dfrac{X}{1000}$
STEP 17   STOP: $\sqrt{A}$

Exercise 3.1
STEP 1   INPUT N
STEP 2 $X = S(1)$
STEP 3   FOR $Y = 2$ to N
STEP 4   IF $X < S(Y)$ THEN $X = S(Y)$
STEP 5   STOP

Exercise 3.2   4, 32;   Pick powers of 2 from a list.

Exercise 3.3

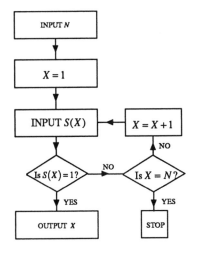

Exercise 3.4 (a) area = 6
(b) 84, 306, no such triangle, 3069 the third 'triangle' cannot be drawn.

Exercise 3.5 (a) 4, 8, 3, 9   (b) 2, 8, 11, 3, 7, 9
The numbers are re-arranged so all the even ones are listed before the odd ones.

Exercise 3.6 (a) and (c)

Exercise 4.1 It takes a maximum of $n + 1$ steps to go round the feasible region to find the optimal vertex. $0(n)$ as there are at most $2n + 2$ nodes so $2n + 1$ steps.

Exercise 4.2 | 1 | 5 | 10 | 20 | 50 | 100 |
|---|---|---|---|---|---|
| 2×10^{-6} | 3.2×10^{-5} | 1.024×10^{-3} | 1 sec | 36 years | 4×10^{16} years |

Inefficient!

Exercise 4.3 A is efficient, B and C are inefficient.

Exercise 4.4 n

Index